COMPLEX VARIABLES

BY

H. R. CHILLINGWORTH

College of St. Mark and St. John
King's Road, Chelsea

PERGAMON PRESS

OXFORD · NEW YORK · TORONTO
SYDNEY · BRAUNSCHWEIG

Pergamon Press Ltd., Headington Hill Hall, Oxford
Pergamon Press Inc., Maxwell House, Fairview Park, Elmsford,
New York 10523
Pergamon of Canada Ltd., 307 Queen's Quay West, Toronto 1
Pergamon Press (Aust.) Pty. Ltd., 19a Boundary Street,
Rushcutters Bay, N.S.W. 2011, Australia
Vieweg & Sohn GmbH, Burgplatz 1, Braunschweig

First edition 1973
Library of Congress Catalog Card No. 72-86178

Printed in Hungary

$A / 517.8$

This book is sold subject to the condition
that it shall not, by way of trade, be lent,
resold, hired out, or otherwise disposed
of without the publisher's consent,
in any form of binding or cover
other than that in which
it is published.

08 016938 4 (Hardcover)
08 016939 2 (Flexicover)

THE COMMONWEALTH AND INTERNATIONAL LIBRARY

Joint Chairmen of the Honorary Editorial Advisory Board

SIR ROBERT ROBINSON, O.M., F.R.S., LONDON

DEAN ATHELSTAN SPILHAUS, MINNESOTA

MATHEMATICAL TOPICS

General Editor: C. PLUMPTON

COMPLEX VARIABLES

CONTENTS

INTRODUCTION

THE material in this book covers the syllabuses of the complex variable sections of the London B.Ed. degree, and thus, incidentally, most of the topics in the corresponding section of the London B.Sc.

Being written mainly for intending teachers, the book first introduces complex numbers in a manner which I consider to be valuable from an educational viewpoint—via the selection, from an at first apparently unrestricted choice, of a set of simple and convenient rules of manipulation based on geometrical considerations, leaving the reader, if he will, to experiment with alternative sets of rules. (Several of the exercises are deliberately "open-ended"). Two other approaches in which geometrical ideas are replaced by algebraic definitions from the start (in keeping with the prevailing taste for maximum rigour— often at the expense of intuition) are also outlined. I am, however, not in favour of these methods of introduction. I regard the tendency not to distinguish between the concepts of a point in a plane and an ordered number pair, on the grounds that geometrical concepts (with their associated philosophical problems) are so much harder to handle than algebraic ones that it is better to ignore the former completely, as in the long run quite pernicious—however satisfying the formal manipulation of symbols may indeed be in its own right.

At a time when complex variable is regarded as a "dead" subject in comparison with algebra and topology, some justification should perhaps be given for my conviction that the subject is of real value to teachers of mathematics. Some of its merits are seen as follows:

(1) It is possible to present it as a natural development of the real number system with rules of manipulation selected (with due attention to consistency) merely for their intrinsic appeal; the exercise of taste

in the construction of such a system is a valuable mathematical experience.

(2) The system provides an example of a number field in which the field of real numbers is properly embedded. The generalization of the idea of number, with the suggestion of further possibilities (quaternions, vectors, matrices), can be stimulating to an imaginative student.

(3) Complex numbers are a powerful unifying concept in elementary algebra since after they are introduced it is possible to state that all polynomial equations of degree n have exactly n roots. In particular, all quadratic equations with real (complex) coefficients become completely soluble.

(4) The "pay-off" comes more quickly than with many subjects in which so much of the early work is merely definition and description, and in which significant results take a long time to reach.

(5) Complex numbers are a very useful tool in applied subjects such as hydrodynamics and electricity as well as in trigonometry, differential equations, algebraic geometry, number theory, etc.

(6) Parts of the subject have a particular aesthetic appeal—notably the simplifying light shed upon certain results in real analysis and the elegant method of contour integration based on the theory of residues.

(7) It provides a natural framework within which to introduce some elementary concepts from topology (neighbourhoods, open sets, etc.).

Exercises have been included for every section of each chapter except the last; but to avoid unduly breaking up the text these have been placed at the end of each chapter. The system of enumeration (3.5.2, for example, referring to the second exercise relating to section 5 of chapter 3) should be clear.

My indebtedness to the Hardy–Titchmarsh line of development of the subject will be too obvious to need further specific reference. I am grateful to the University of London for their permission to make use of examination questions set in the B.Sc. and B.Ed. examinations. These are included as miscellaneous examples at the end of the book. I am specially indebted to my son David who has made many valuable suggestions and has also drawn my attention to quite a few errors—those that remain being, of course, all my own.

Finally, my thanks are due to the publishers for their helpfulness and consideration.

H. R. CHILLINGWORTH

CHAPTER 1

COMPLEX NUMBERS

THE theory of complex numbers may be developed either from an algebraic or from a geometrical viewpoint. We shall examine both approaches. Since, however, geometrical considerations will have a large part to play in much of the theory that follows, it is these that we shall emphasize in the first sections.

1.1. The Complex Plane

In the well-known representation of real numbers by means of points on a "number line" the line is by practical necessity depicted as lying in a plane—the "plane of the paper". This situation leads us to consider those points of the plane which do not lie on the line and to the possibility of taking them to represent "numbers". They cannot usefully be regarded as representing real numbers since all of these are already accounted for by points on the line. On the other hand, there is nothing to prevent us, if we so wish, from regarding them as representing an entirely new type of "number", and this is what we shall proceed to do. Let us start by expressing our ideas more precisely.

In a plane p we take a straight line l, to represent the real numbers; thus on l we have a point 0 representing the number zero. The numbers represented by points on one side of 0, which we will regard as the "right-hand side", are called positive, and those represented by points on the other side (the "left-hand side") are called negative, and we think of the number x represented by a point X on l as the distance from 0 to X if X is on the right-hand side of 0, or minus the distance if X is on the left-hand side. The distance OX is called the modulus of x, denoted as usual by $|x|$.

Now all the points in the plane p, whether on the line l or not, will be taken as representing our new "numbers". For the present we shall refer to these new "numbers" as p-numbers (p = plane). The properties of these numbers are largely at our disposal—we are free to assign to them any properties we may wish so long as these are not mutually contradictory; on the other hand, we shall hope to be able to produce an interesting and non-trivial mathematical structure, capable of further development, from the properties that we postulate. One such set of properties is found to be of particular importance. The p-numbers with these properties (which we shall develop later) are called *complex numbers*, and the plane is in this case referred to as the *complex plane*.

1.2. Modulus

The real numbers form a totally ordered set under the relation "is less than". That is to say, if we are given two different real numbers a and b we can always assert either that $a < b$ or that $b < a$. How to define which is the "larger" of two p-numbers presents a problem. We attempt to treat this, by analogy with positive real numbers, by considering the distance of the point representing a given p-number from the point 0 representing the number zero on l. We then agree that the "larger" the number the greater will be the distance from 0 of the point representing it. We call this distance the *modulus* of the p-number and regard it as measuring the "magnitude" of the number. We immediately note that in general there are an infinite number of p-numbers with the same magnitude r—namely the numbers corresponding to all the points on the circle with centre 0 and radius r. (The single exception is given by $r = 0$. Thus the only p-number of zero magnitude is represented by the same point which represents the real number zero.) Of the p-numbers of modulus $r > 0$, two have representative points on the real line and correspond to the real numbers r and $-r$.

Since we wish to regard all the points of the plane as representing *p*-numbers, we have to include—as *p*-numbers—the points on *l* all of which already represent real numbers; hence, when selecting properties for *p*-numbers we must try to make sure that there is no inconsistency between regarding points on the real number line as (i) real numbers, or (ii) *p*-numbers.

So far this has applied; our definition of modulus coincides for real numbers with the previous definition (see § 1.1). We shall use the same notation also; if *z* is a *p*-number with modulus *r*, we write $r = |z|$.

1.3. Amplitude

The modulus of a *p*-number is obviously insufficient to determine the position of its representative point. More information is needed. The natural suggestion is to use the angle made with the line *l* by the join of 0 to the representative point as follows.

If a ray through 0 along *l* in the positive direction is rotated counterclockwise, we regard the angle turned through as positive; clockwise rotation gives negative angles. If *P* is a point of *p* other than 0, then the angle θ, such that $-\pi < \theta \leqslant \pi$, which determines the direction of the ray 0*P*, will be called the *principal amplitude*† of the *p*-number represented by *P*. It will be observed that adding $2n\pi$ ($n = \pm 1, \pm 2, \ldots$) to the principal amplitude will not alter the direction of the ray, and therefore any angle so obtained may be taken instead and called an *amplitude*.† The term "principal amplitude" is reserved for that amplitude which falls within the range stated ($-\pi < \theta \leqslant \pi$). It is clear from the above that a *p*-number is uniquely determined by its modulus together with any one of its amplitudes.

From now on, in order to avoid undue pedantry, we shall frequently refer to the points of the plane themselves as *p*-numbers, or, later, complex numbers rather than as points representing such numbers. This is in the same way that we refer, for example, to "the point -3" on the real number line. We shall also occasionally regard the position vector **0P** as representing the *p*-number associated with *P*. Thus a *p*-number may be, according to convenience, regarded as:

† Or, *argument*.

(i) a position vector,

(ii) a point,

(iii) a mathematical entity with properties related to these represen-
tations.

1.4. Number Pairs

When the modulus and amplitude of a p-number are given, the
number is uniquely determined. These are not, however, the only two
measurements available for this purpose, and we now consider the
alternative way of expressing the position of a p-number using the
more familiar cartesian coordinates. In fact, by taking an axis l_1
through 0 perpendicular to l we can locate any p-number by means
of the ordered pair (x, y), where x denotes a measurement along l
(the "x-axis") from 0, and y a measurement perpendicular to it, or
parallel to l_1 (the "y-axis"). Positive or negative signs are attached in
the usual way to these real numbers x, y to indicate whether displacements are to the right or the left of 0 and whether the points are
"above" or "below" the x-axis. We refer to the x-axis, as the *real axis*
and the y-axis as the *imaginary axis.*

DEFINITION. *If z is the p-number represented by the point with
coordinates (x, y), then the number x is called the* real part *of z and the
number y is called the* imaginary part *of z. A p-number with coordinates
$(x, 0)$ is referred to as* real *and identified with the real number x, while
a number with coordinates $(0, y)$ (corresponding to a point on the y-axis)
is referred to as* imaginary. *The p-number with coordinates $(0, 0)$ is
called* zero, *and identified with the real number zero on l.*

We note the following important relationships:

$$x = r \cos \theta,$$
$$y = r \sin \theta,$$
$$r = +\sqrt{(x^2+y^2)},$$
$$\cos \theta = x/r,$$
$$\sin \theta = y/r.$$

Another very useful relationship is $\tan \theta = y/x$, but it must be
noted that this equation will give two values for θ, differing by π,

in the range $-\pi < \theta \leqslant \pi$, only one of which will be the amplitude of (x, y). For example $(-1, 1)$ and $(1, -1)$ have principal amplitudes of $3\pi/4$ and $-\pi/4$ respectively, but in each case $\tan \theta = -1$. (Recourse to a diagram is advisable in case of difficulty.) (Figs. 1.1 and 1.2.)

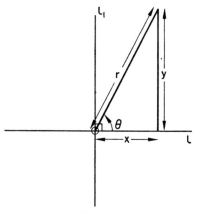

FIG. 1.1.

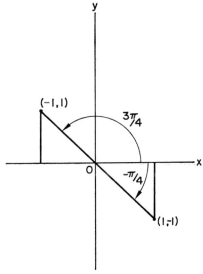

FIG. 1.2.

1.5. Addition

We now turn to the problem of inventing a consistent arithmetic for these p-numbers. To begin with, how shall we add two p-numbers together? More precisely, how shall we add a second p-number to a given p-number? When trying to construct a rule we must bear in mind there is no *a priori* necessity to require that if a and b are two p-numbers, then $a+b = b+a$; but such a result would certainly be desirable since it already applies to real numbers, and we naturally wish it to continue to do so when they are treated as p-numbers.

In illustrating by means of the real-number line the addition of b to a, when a and b are real numbers, we lay a line segment representing b in magnitude and direction "end to end" with the line segment representing a and originating from 0, and then find the location of the free end point of the b segment, i.e. we treat the problem as an elementary case of the addition of two vectors. It seems reasonable to adopt the same procedure of "vector addition" for p-numbers, and this is exactly what we shall do (Fig. 1.3).

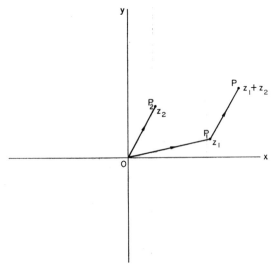

Fig. 1.3. Addition of p-numbers.

If P_1 and P_2 represent the p-numbers z_1 and z_2, we construct a line segment P_1P equal in length and parallel to $0P_2$, and define z_1+z_2 to be the p-number represented by P. It is then easily shown (see Exercise 1.5.1) that:

(a) if, using cartesian coordinates, $z_1 = (x_1, y_1)$ and $z_2 = (x_2, y_2)$, then $z_1+z_2 = (x_1+x_2, y_1+y_2)$;

(b) $z_1+z_2 = z_2+z_1$ (*the commutative law for addition*);

(c) $z_1+(z_2+z_3) = (z_1+z_2)+z_3$ (*the associative law for addition*).

Since real numbers obey the commutative and associative laws, and we are attempting to find rules of operation for p-numbers which follow as far as possible the same pattern as those for real numbers, the fact that p-numbers obey these laws (for addition) is so far very satisfactory.

1.6. Scalar Multiplication

We note that if $z = (x, y)$ then $z+z = (2x, 2y)$, $z+z+z = (3x, 3y)$, and so on. Evidently we could agree to write these relationships in the form $2z = (2x, 2y)$, $3z = (3x, 3y)$, $\ldots$, $kz = (kx, ky)$, where k is any positive integer. We have here *defined* kz to mean $z+z+\ldots$ to k terms. We now generalize this rule, writing $a(x, y)$ to mean (ax, ay), where a is any real number whatever.

Two particular cases are of special importance. One is

$$0(x, y) = (0, 0),$$

whence any p-number multiplied by zero gives zero. The other is

$$-1(x, y) = (-x, -y)$$

which we can write as $-1z = -z$, as with real numbers; we have here *defined* $-z$ to mean $(-x, -y)$, where $z = (x, y)$. The symbol $-z$ would otherwise be meaningless at this stage.

We note also that

$$1(x, y) = (x, y), \quad \text{or} \quad 1\cdot z = z.$$

1.7. Subtraction

Having defined addition, we have not a similar choice of definition
with regard to subtraction. In fact we require subtraction to be the
operation inverse to addition. More precisely, we require that z_3-z_2
shall mean the number which added to z_2 will give z_3. From this relation it is easy to see that the only definition possible is that given by
the rule

if $\ z_1 = (x_1, y_1)\ $ and $\ z_2 = (x_2, y_2)\ $ then $\ z_1-z_2 = (x_1-x_2, y_1-y_2)$

or

$$z_1-z_2 = z_1+(-1)z_2 = z_1+(-z_2).$$

In particular

$$z_1+(-z_1) = z_1-z_1 = (0, 0) = 0.$$

Subtraction is easily interpreted geometrically; to subtract z it is
merely necessary to add $-z$, as already defined. In examining the
geometrical interpretation it is useful to note that if the points corre-

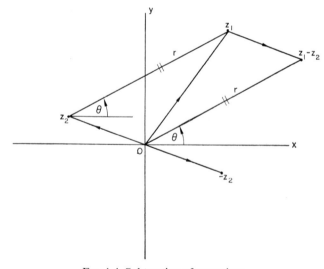

Fig. 1.4. Subtraction of p-numbers.

sponding to z_1 and z_2 are plotted on the plane p, the directed line-segment from z_2 to z_1 represents in magnitude and direction the position vector corresponding to $(z_1 - z_2)$, so that the modulus and amplitude of $z_1 - z_2$ are available in the diagram without it being necessary to plot the actual point corresponding to $z_1 - z_2$. (Fig. 1.4.)

1.8. Multiplication

We have already defined what we mean by the multiplication of a p-number by a real number. Since we wish the same result to be produced whether we regard the real number as such or as a p-number, we are committed to the requirement that

$$(x, 0) \times (x_1, y_1) = (xx_1, yy_1)$$

and thus, rather surprisingly perhaps, we are unable to adopt what may seem the "obvious" rule

$$(x_1, y_1) \times (x_2, y_2) = (x_1 x_2, y_1 y_2)$$

for multiplication of p-numbers. In any case, this rule would have been rather unsatisfactory, since it would involve the conclusion that $(x, 0) \times (0, y) = (0, 0) = 0$, and we would not wish the product of two non-zero numbers to be zero if this could be avoided.

A guide towards a suitable rule is provided by the following observation. When two real numbers are added, the magnitude of the result is not necessarily the sum of the magnitudes of the two separate components (although it may be). For example,

$$(-5) + (+2) = (-3).$$

Here the sum of the moduli of the separate numbers on the left is $+7$ while the modulus of their sum is $+3$. On the other hand, when two real numbers are multiplied together, the result is always of the "right size":

$$(-5) \times (+2) = (-10),$$
$$(-5) \times (-2) = (+10),$$

and in general the modulus of the product is the product of the moduli of the factors. We shall try to devise a rule for the multiplication of p-numbers which will give us a product with the "correct" modulus, explicitly; we shall look for a rule such that, for any two numbers $z_1, z_2,$

$$|z_1 \times z_2| = |z_1| \times |z_2|. \quad \text{(See also Appendix A.)}$$

Let us write z_1, z_2 in the form

$$z_1 = (r_1 \cos \theta_1, r_1 \sin \theta_1), \quad z_2 = (r_2 \cos \theta_2, r_2 \sin \theta_2);$$

then we evidently require a rule of the form

$$z_1 \times z_2 = (r_1 \cos \theta_1, r_1 \sin \theta_1) \times (r_2 \cos \theta_2, r_2 \sin \theta_2)$$
$$= (r_1 r_2 \cos \phi, r_1 r_2 \sin \phi),$$

where ϕ is some angle still to be determined, by applying some suitable rule to θ_1 and θ_2. Of course we could devise some more complicated type of rule in which the angle ϕ depended also on r_1 and r_2, but we shall try to preserve simplicity, so, at least initially, we shall attempt to find some simple way of combining θ_1 and θ_2 to give us an acceptable ϕ.

It will not be satisfactory simply to multiply θ_1 by θ_2. This rule would give, for example,

$$(-2, 0) \times (-3, 0) = (2 \cos \pi, 2 \sin \pi) \times (3 \cos \pi, 3 \sin \pi)$$
$$= (6 \cos \pi^2, 6 \sin \pi^2)$$
$$\neq (6, 0), \text{ the required answer.}$$

Suppose we examine addition of the angles. In the case considered this would give

$$(-2, 0) \times (-3, 0) = (2 \cos \pi, 2 \sin \pi) \times (3 \cos \pi, 3 \sin \pi)$$
$$= (6 \cos 2\pi, 6 \sin 2\pi)$$
$$= (6, 0), \text{ which is satisfactory.}$$

More generally, if a and b are positive real numbers, we have

(i) $(a, 0) \times (b, 0) = (a \cos 0, a \sin 0) \times (b \cos 0, b \sin 0)$
$$= (ab \cos 0, ab \sin 0)$$
$$= (ab, 0).$$

(ii) $(-a, 0) \times (b, 0) = (a \cos \pi, a \sin \pi) \times (b \cos 0, b \sin 0)$
$$= (ab \cos \pi, ab \sin \pi)$$
$$= (-ab, 0).$$

(iii) $(a, 0) \times (-b, 0) = (a \cos 0, a \sin 0) \times (b \cos \pi, b \sin \pi)$
$$= (ab \cos \pi, ab \cos \pi)$$
$$= (-ab, 0).$$

(iv) $(-a, 0) \times (-b, 0) = (a \cos \pi, a \sin \pi) \times (b \cos \pi, b \sin \pi)$
$$= (ab \cos 2\pi, ab \sin 2\pi)$$
$$= (ab, 0).$$

All of these results are in accordance with those obtained by interpreting the left-hand side in each case as the product of real numbers.

We also need to check the result given by this method of multiplication (multiplication of moduli and addition of amplitudes) in evaluating such products as $(x, 0) \times (x_1, y_1)$, since we have above already defined this product to be (xx_1, yy_1). (See §§ 1.8 and 1.6.)

Writing $(x_1, y_1) = (r_1 \cos \theta, r_1 \sin \theta)$, we have, if $x > 0$,

$(x, 0) \times (x_1, y_1) = (x \cos 0, x \sin 0) \times (r_1 \cos \theta, r_1 \sin \theta)$
$$= (xr_1 \cos \theta, xr_1 \sin \theta)$$
$$= (xx_1, yy_1)$$

and if $x < 0$,

$(x, 0) \times (x_1, y_1) = (-x \cos \pi, -x \sin \pi) \times (r_1 \cos \theta, r_1 \sin \theta)$
$$= [-xr_1 \cos (\pi + \theta), -xr_1 \sin (\pi + \theta)]$$
$$= (xr_1 \cos \theta, xr_1 \sin \theta)$$
$$= (xx_1, yy_1) \quad \text{as required.}$$

We have thus now checked that our rule for defining $z_1 \times z_2$ is consistent with the rule we had already defined in the case when z_1 is real. Hence the rule seems to be acceptable, and we shall provisionally adopt it; i.e. if $z_1 = (r_1 \cos \theta_1, r_1 \sin \theta_1)$ and $z_2 = (r_2 \cos \theta_2, r_2 \sin \theta_2)$, then

$$z_1 \times z_2 = [r_1 r_2 \cos (\theta_1 + \theta_2), r_1 r_2 \sin (\theta_1 + \theta_2)].$$

It must be noted, however, that if we are using the principal amplitudes of the two numbers we wish to multiply, the amplitude of the product, as given by the above rule, need not be its *principal* amplitude. Look, for example, at one of the cases already considered, viz. $(-a, 0) \times (-b, 0)$, $(a, b > 0)$. The rule we are using gives here an amplitude of 2π (see (iv) above), which is not the principal amplitude. The principal amplitude for a positive real number is zero, not 2π.

With this reservation in mind, we repeat that our rule for multiplication may be expressed simply as "multiply the moduli; add the amplitudes".

We shall often write $z_1 \times z_2$ in the form $z_1 z_2$, as with real numbers. Turning again to the cartesian form for p-numbers, we have

$$(x_1, y_1) \times (x_2, y_2) = (r_1 \cos \theta_1, r_1 \sin \theta_1) \times (r_2 \cos \theta_2, r_2 \sin \theta_2), \text{ say,}$$
$$= [r_1 r_2 \cos (\theta_1 + \theta_2), r_1 r_2 \sin (\theta_1 + \theta_2)],$$

which, on expansion,

$$= (r_1 r_2 \cos \theta_1 \cos \theta_2 - r_1 r_2 \sin \theta_1 \sin \theta_2, r_1 r_2 \sin \theta_1 \cos \theta_2$$
$$+ r_1 r_2 \cos \theta_1 \sin \theta_2),$$

or, on regrouping the multiple products,

$$= (r_1 \cos \theta_1 \, r_2 \cos \theta_2 - r_1 \sin \theta_1 \, r_2 \sin \theta_2, r_1 \sin \theta_1 \, r_2 \cos \theta_2$$
$$+ r_1 \cos \theta_1 \, r_2 \sin \theta_2),$$
$$= (x_1 x_2 - y_1 y_2, y_1 x_2 + x_1 y_2).$$

It may easily be seen from this formula (but is even more easily seen from the "modulus, amplitude" approach) that multiplication as we have defined it obeys the commutative and associative laws for multiplication, namely

$$z_1z_2 = z_2z_1 \quad \text{(the commutative law for multiplication)},$$

$$z_1(z_2z_3) = (z_1z_2)z_3 \quad \text{(the associative law for multiplication)}.$$

We note in particular that the following results are immediately obtained from the cartesian form of the product, which, in view of its importance, we repeat here (slightly rearranged):

$$(x_1, y_1) \times (x_2, y_2) = (x_1x_2 - y_1y_2, \ x_1y_2 + x_2x_1),$$

(i) $(a, 0) \times (b, 0) = (ab, 0)$, the "right" result for real numbers;

(ii) $(a, 0) \times (x, y) = (ax, ay) = a(x, y)$ corresponding to scalar multiplication;

(iii) $(x, y)a = (x, y) \times (a, 0) = (xa, ya) = a(x, y)$;

(iv) [a special case of (ii)]. $0(x, y) = (0, 0) \times (x, y) = (0, 0) = 0$.

1.9. Division

Having defined multiplication, we have no choice with regard to the rule for division, which is derived as follows. We shall wish to define $z_1 \div z_2$ to mean "the p-number z_3 by which z_2 must be multiplied to give z_1". In other words, if $z_1 \div z_2 = z_3$, we shall require that $z_1 = z_3 \times z_2$. This leads to the result (see Exercise 1.9.1)

$$(x_1, y_1) \div (x_2, y_2) = \left[\frac{x_1x_2 + y_1y_2}{x_2^2 + y_2^2}, \ \frac{y_1x_2 - x_1y_2}{x_2^2 + y_2^2} \right] \quad (x_2^2 + y_2^2 \neq 0).$$

We shall usually write $z_1 \div z_2$ as z_1/z_2.

The rules of operation developed above in §§ 1.5–1.9 will now be adopted as our rules for the arithmetic of p-numbers. The p-numbers, together with these rules, will be re-named *complex numbers*, and the plane p will be called the *complex plane*[†]. Using the cartesian coordinate

† Or, *Argand plane*.

form for complex numbers, we can now summarize their arithmetical properties as follows:

Equality: $(x, y) = (u, v)$ if and only if $x = u$, $y = v$.

Addition: $(x, y) + (u, v) = (x+u, y+v)$.

Scalar multiplication: $k(x, y) = (kx, ky)$.

Multiplication: $(x, y) \times (u, v) = (xu-yv, xv+yu)$.

As consequences of the rules for addition and multiplication we have:

Subtraction: $(x, y) - (u, v) = (x-u, y-v)$.

Division: $(x, y) \div (u, v) = [(xu+yv)/(u^2+v^2), (yu-vx)/(u^2+v^2)]$,
$(u, v) \neq (0, 0)$.

1.10. An Alternative Notation

A result of particular significance in the multiplication of complex numbers is the following:

$$(0, 1) \times (0, 1) = (-1, 0).$$

Now $(-1, 0)$ is identified with the real number -1; so we have found a complex number satisfying the equation $z^2 = -1$. To decide whether there may be other complex numbers satisfying this equation, consider $(x, y) \times (x, y) = (-1, 0)$, which gives, on expansion,

$$(x^2-y^2, 2xy) = (-1, 0).$$

This relationship cannot be satisfied unless either x or y is zero, since we require $2xy = 0$. If $x = 0$, then $y^2 = 1$ and $y = \pm 1$, while if $y = 0$, then $x^2 = -1$, which is not possible for any real x. Hence there are exactly two "square roots of -1" among the complex numbers, namely $(0, 1)$ and $(0, -1)$. We notice that $(0, -1)$ is equal to $-(0, 1)$ (by scalar multiplication of $(0, 1)$ by -1). We denote $(0, 1)$ by i, and $(0, -1)$ by $-i$, and then we have $i^2 = (-i)^2 = -1$.

We next note that any complex number $z = (a, b)$ may be written in the form

$$z = (a, 0) + (0, b)$$
$$= a(1, 0) + b(0, 1)$$
$$= a + bi, \quad \text{or} \quad a + ib.$$

Therefore if we denote the real part of z (see § 1.4) by $\mathcal{R}(z)$ and the imaginary part by $\mathcal{I}(z)$ we see that we can express any complex number z as

$$z = \mathcal{R}(z) + i\mathcal{I}(z).$$

If $\mathcal{I}(z) = 0$, then z is a real number, and if $\mathcal{R}(z) = 0$, then z is called an *imaginary* number. If $\mathcal{I}(z) > 0$, then z is said to lie in the *upper half-plane*.

The $x + iy$ notation presents great advantages for the simple algebraic manipulation of complex numbers. For example, if we merely treat i as a new symbol appearing in combinations with the real numbers $a, b, c, \ldots, k \ldots$, we may write the results we have already obtained in the forms below:

$$(a + ib) \pm (c + id) = (a \pm c) + i(b \pm d),$$
$$(a + ib)(c + id) = ac + aid + ibc + i^2bd$$
$$= ac - bd + i(ad + bc) \quad (\text{since } i^2 = -1),$$
$$(a + ib)/(c + id) = (a + ib)(c - id)/(c + id)(c - id)$$
$$= [ac + bd + i(bc - ad)]/(c^2 + d^2). \quad (c^2 + d^2) \neq 0.$$

1.11. An Algebraic Approach

The definition of complex numbers which we have arrived at above was based on the geometry of the plane. However, it is possible to define complex numbers in a way which is purely algebraic and independent of the geometry of the plane, as follows.

We *define* a complex number to be merely an ordered pair (a, b) of real numbers, with rules of arithmetic defined as follows:

(1) The complex numbers $(a, b), (c, d)$ are *equal* if and only if $a = c$ and $b = d$.

(2) The *sum* $(a, b) + (c, d)$ is *defined* to be $(a + c, b + d)$.

From (2) we see that $(a, b)+(0, 0) = (a+0, b+0) = (a, b)$, so the addition of $(0, 0)$ to any complex number does not alter it. We therefore call $(0, 0)$ the (complex number) *zero*. Also we have $(a, b)+(-a, -b)$ $= (a-a, b-b) = (0, 0)$, and we call $(-a, -b)$ the *additive inverse* of (a, b), and denote it by $-(a, b)$. *Subtraction* is then *defined* by

(2′) $(a, b)-(c, d) = (a, b)+[-(c, d)]$.

Note the following features of the set of all ordered pairs of real numbers together with the above rule of addition:

   (a) the operation of addition obeys the associative law (check this);
   (b) there is a *unique* element (the zero) which when added to any other element does not change it (we call such an element an "identity element" for addition);
   (c) for every element (a, b) there exists a unique element (its additive inverse) which when added to (a, b) gives the identity element.

The fact that (a), (b), and (c) hold means that the set of complex numbers with addition as defined forms a *group* (see Appendix B). We also have

   (d) addition obeys the commutative law,

and so this group is, in fact, an *abelian group*.

(3) *Multiplication* of complex numbers (real number pairs) is next defined by the rule

$$(a, b)\times(c, d) = (ac-bd, ad+bc).$$

Since, according to these rules,

$$(a, 0)+(b, 0) = (a+b, 0)$$

and

$$(a, 0)\times(b, 0) = (ab, 0),$$

we can clearly identify a number of the form $(a, 0)$ with the corresponding real number a. We then *deduce*

$$a(x, y) = (a, 0)\times(x, y) = (ax, ay) \text{ for any real number } a,$$

which was a result we *assumed* in our original approach when considering possible definitions of multiplication of "p-numbers". In particular $0(x, y) = (0, 0) = 0$.

Division is now treated as the inverse of multiplication. We proceed as follows. For every (x, y) except $(0, 0)$ we can find a unique (u, v) such that $(x, y) \times (u, v) = (1, 0) = 1$. In fact, from the equations

$$xu - yv = 1,$$
$$xv + yu = 0,$$

we immediately obtain the results

$$u = x/(x^2 + y^2), \quad v = -y(x^2 + y^2).$$

The number pair (u, v) is called the multiplicative inverse of (x, y). In order to divide by a complex number we multiply by its multiplicative inverse. This gives, for example,

$$
\begin{aligned}
(3') \quad (p, q) \div (x, y) &= (p, q) \times (u, v) \\
&= (p, q) \times [x/(x^2 + y^2), \, -y/(x^2 + y^2)] \\
&= [(px + qy)/(x^2 + y^2), \, (qx - py)/(x^2 + y^2)].
\end{aligned}
$$

Thus we have now arrived, by an entirely different route, at the same set of operations as before. The approach—particularly as regards multiplication—seems highly arbitrary, since there appears to be no reason whatsoever to define multiplication in this complicated fashion. On the other hand, by avoiding any appeal to intuitive ideas we are able to establish complex number theory on a sound axiomatic basis depending only upon real numbers.

Note that if we exclude the additive identity $(0, 0)$, number-pairs form an abelian group under multiplication, the multiplicative identity element being $(1, 0)$. In addition to this, multiplication as so defined is distributive over addition, i.e. if a, b, and c represent complex numbers, then

$$a(b + c) = ab + ac,$$
$$(b + c)a = ba + ca.$$

A system possessing all these properties—being an abelian group under both addition and (when the additive identity is excluded) multiplication, and with multiplication distributive over addition—is called a *field*. Thus the complex numbers form a field, and it may easily be verified that the real numbers form another field contained within it [we identify $(a, 0)$ with a].

In this development of complex numbers we have not so far met either of the important concepts *modulus* and *amplitude*. These are introduced entirely artificially, *defining* the *modulus* of (x, y) as $+\sqrt{(x^2+y^2)}=r$, say; and *any* angle θ such that $\cos \theta=x/r$, $\sin \theta=y/r$, as an *amplitude*—reserving the term *principal amplitude* for that angle θ which satisfies $-\pi < \theta \leqslant \pi$.

We have now reached the point of development that we had originally reached at the end of § 1.9. As in § 1.10, we may adopt the alternative notation of writing i for $(0, 1)$, and so on.

1.12. Complex Numbers as an Extension of the Real Number Field

Another algebraic approach to complex numbers which we outline here, is motivated by the observation that the equation $x^2+1 = 0$ has no real roots. We *define* a new "number" i, and construct a field F whose elements are expressions of the form

$$\frac{a_0+a_1i+a_2i^2+ \ldots +a_ni^n}{b_0+b_1i+b_2i^2+ \ldots +b_mi^m} = \frac{P}{Q}, \quad \text{say,}$$

where the a's and b's are real numbers. Addition and multiplication are defined in the obvious way:

$$\frac{P}{Q}+\frac{R}{S} = \frac{PS+QR}{QS}, \quad \frac{P}{Q}\times\frac{R}{S} = \frac{PR}{QS}.$$

We then postulate that whenever i^2 appears we shall replace it by -1. Every element of the field F can then be reduced to an element of form

$$\frac{c_0+c_1i}{d_0+d_1i},$$

which, by multiplying top and bottom by $d_0 - d_1 i$,† reduces to

$$\frac{c_0 d_0 + c_1 d_1}{d_0^2 + d_1^2} + \frac{(c_1 d_0 - c_0 d_1)i}{d_0^2 + d_1^2} = x + yi \quad \text{say.}$$

It is now easy to check that the rules for addition and multiplication for these elements of F are exactly the same as the rules for addition and multiplication of complex numbers as defined in §§ 1.5–1.10 or in § 1.11. Modulus and amplitude can be defined just as in § 1.11.

It is interesting at this point to note that we are able to provide, by means of the additional symbol i, solutions not only to the equation $z^2 + 1 = 0$, but also to every equation of the form $az^2 + bz + c = 0$ with a, b, and c real. This is because if $b^2 < 4ac$ we have $[i\sqrt{(4ac - b^2)}]^2 = (-1)(4ac - b^2) = b^2 - 4ac$, and so the roots are in fact

$$\frac{-b \pm \sqrt{(b^2 - 4ac)}}{2a}, \ (b^2 \geqslant 4ac), \quad \text{or} \quad \frac{-b}{2a} \pm \frac{i\sqrt{(4ac - b^2)}}{2a}, \ (b^2 < 4ac).$$

We find later that the stipulation that a, b, and c be real is unnecessary; solutions still exist if they are complex. If, however, they are real, the roots are of form $x + iy$, $x - iy$, and complex numbers related in this way are of considerable importance in the theory.

1.13. Complex Conjugates

The number $x - iy$ is said to be *conjugate* to $x + iy$. Note that it then follows that $x + iy = x - i(-y)$ is conjugate to $x - iy$. Thus two complex numbers are conjugate precisely when their real parts are the same and their imaginary parts equal and opposite. If $z = x + iy$ we write $\bar{z} = x - iy$. Evidently $\bar{\bar{z}} = z$, since changing the sign of the imaginary part twice will re-establish the *status quo*; and $z, \bar{z}$ are each conjugate to the other. If, for a given z, $z = \bar{z}$, this can only be because $y = 0$, so *a complex number is equal to its conjugate if and only if it is real*. Complex conjugates have equal moduli ($|z| = |\bar{z}|$) and principal amplitudes of opposite signs (with the exception of negative real numbers; if $a > 0$, the principal amplitude of $-\bar{a} = -a$ is π and not

† We here assume that d_0 and d_1 are not both zero.

A/577.8

$-\pi$). The points representing them in the complex plane may each be regarded as the reflection of the other in the real axis. (Check these statements.)

The following results are very important. The proofs, which are extremely easy, are set as exercises (1.13.1):

$$z\bar{z} = x^2 + y^2 = r^2, \quad \text{where} \quad r = |z|.$$
$$\overline{z_1 + z_2} = \bar{z}_1 + \bar{z}_2,$$
$$\overline{z_1 z_2} = \bar{z}_1 \bar{z}_2,$$
$$\overline{(z_1/z_2)} = \bar{z}_1/\bar{z}_2$$
$$\frac{z+\bar{z}}{2} = \mathcal{R}(z), \quad \frac{z-\bar{z}}{2i} = \mathcal{I}(z)$$

$|\mathcal{R}(z)| \leqslant |z|, |\mathcal{I}(z)| \leqslant |z|$, equality only occurring in the first case for real numbers and in the second case for imaginary numbers.

As an example of the use of the conjugate in calculations, we will verify a result which we assumed in our geometrical approach, namely that $|z_1 z_2| = |z_1| \times |z_2|$.

We have, in fact,

$$|z_1 z_2|^2 = (z_1 z_2)\overline{(z_1 z_2)}$$
$$= (z_1 z_2)(\bar{z}_1 \bar{z}_2)$$
$$= (z_1 \bar{z}_1)(z_2 \bar{z}_2)$$
$$= |z_1|^2 |z_2|^2.$$

Since moduli are non-negative, the result follows on taking the square root of each side.

1.14. The Triangle Inequality

It is easy to show geometrically (Exercise 1.14.1) that

$$|z_1 + z_2| \leqslant |z_1| + |z_2|,$$

equality occurring if and only if z_1 and z_2 have the same amplitude, or if one of them is zero. Algebraically we may establish this very important result as follows:

$$\begin{aligned}
|z_1+z_2|^2 = (z_1+z_2)\overline{(z_1+z_2)} &= (z_1+z_2)(\bar{z}_1+\bar{z}_2)\\
&= z_1\bar{z}_1+z_2\bar{z}_2+(z_1\bar{z}_2+\bar{z}_1z_2)\\
&= z_1\bar{z}_1+z_2\bar{z}_2+(z_1\bar{z}_2+\overline{z_1\bar{z}_2})\\
&= |z_1|^2+|z_2|^2+2\mathcal{R}(z_1\bar{z}_2)\\
&\leqslant |z_1|^2+|z_2|^2+2|z_1\bar{z}_2|\\
&= |z_1|^2+|z_2|^2+2|z_1||z_2|\\
&= (|z_1|+|z_2|)^2,
\end{aligned}$$

whence, taking the positive square root of each side, we have

$$|z_1+z_2| \leqslant |z_1|+|z_2|.$$

This result may obviously be extended by induction to prove that

$$|z_1+z_2+ \ldots +z_n| \leqslant |z_1|+|z_2|+ \ldots +|z_n|. \quad \text{(Exercise 1.14.5.)}$$

If in the result just obtained we replace z_1 by z_1-z_2 we obtain

$$|z_1| \leqslant |z_1-z_2|+|z_2|,$$

or

$$|z_1|-|z_2| \leqslant |z_1-z_2|,$$

which is another useful form of the inequality, when $|z_1| > |z_2|$; if $|z_1| \leqslant |z_2|$ the result, although still true, becomes trivial, since in this case the left-hand side is less than or equal to zero, and the right-hand side greater than or equal to zero.

1.15. De Moivre's Theorem

The following powerful theorem has an elementary proof.

THEOREM (De Moivre's theorem). *If n is a positive integer and θ a real number, then*

$$(\cos \theta +i \sin \theta)^n = \cos n\theta +i \sin n\theta.$$

Proof. By induction. The result is trivially true when $n = 1$. Assume it to be true for $n = N$.

Then
$$(\cos \theta + i \sin \theta)^{N+1} = (\cos \theta + i \sin \theta)^N (\cos \theta + i \sin \theta)$$
$$= (\cos N\theta + i \sin N\theta)(\cos \theta + i \sin \theta)$$
$$= \cos N\theta \cos \theta - \sin N\theta \sin \theta + i (\sin N\theta \cos \theta + \cos N\theta \sin \theta)$$
$$= \cos (N+1)\theta + i \sin (N+1)\theta.$$

Hence if the formula is true for N then it is true for $N+1$. Since it is true for 1 it is also true for 2, 3, ..., and for all n. This proves the theorem.

Moreover,

$$(\cos \theta + i \sin \theta)^{-n} = 1/(\cos \theta + i \sin \theta)^n$$
$$= 1/(\cos n\theta + i \sin n\theta), \text{ (by the theorem)}$$
$$= (\cos n\theta - i \sin n\theta)/(\cos^2 n\theta - i^2 \sin^2 n\theta)$$
$$= (\cos n\theta - i \sin n\theta)/(\cos^2 n\theta + \sin^2 n\theta)$$
$$= \cos (-n)\theta + i \sin (-n)\theta,$$
[since $\cos (-A) = \cos A$, $\sin (-A) = -\sin A$].

Hence the theorem is seen to be true also for negative integers, and thus (since it is evidently true for $n = 0$) for all integers.

We now consider $(\cos \theta + i \sin \theta)^{1/n}$, which we interpret as meaning any nth root of the complex number $\cos \theta + i \sin \theta$, i.e. any complex number which when raised to the nth power will give $\cos \theta + i \sin \theta$. We will assume such a number exists, and write it in the form $r (\cos \phi + i \sin \phi)$. By raising to the nth power each side of the equation

$$(\cos \theta + i \sin \theta)^{1/n} = r(\cos \phi + i \sin \phi)$$

we obtain

$$\cos \theta + i \sin \theta = r^n(\cos n\phi + i \sin n\phi)$$

whence, by equating moduli, we see that $r^n = 1$; from which it follows, since r is real and non-negative, that $r = 1$.

We then have $\cos \theta = \cos n\phi$, $\sin \theta = \sin n\phi$, and thus the only possible values of $n\phi$ are of form $\theta + 2k\pi$ $(k = 0, \pm 1, \pm 2, \ldots)$; so we obtain the following possible values for ϕ:

$$\phi = \theta/n, (\theta - 2\pi)/n, (\theta + 2\pi)/n, (\theta - 4\pi)/n, \ldots.$$

In due course we arrive at the value $\phi = (\theta/n - 2n\pi)/n = \theta/n - 2\pi$, giving the same complex number as θ/n; and from this point on, all the values are repetitions of values previously listed. Thus we do not obtain an infinite number of nth roots of $\cos \theta + i \sin \theta$, but in fact precisely n. This is perhaps more easily seen by first considering only positive values for k. The values $k = 0, 1, 2, \ldots, (n-1)$ all give different roots, after which the same roots repeat in the same order. As for negative values, it is merely necessary to observe that for each k listed,

$$\cos (\theta - 2k\pi/n) = \cos [\theta + 2(n-k)\pi/n],$$

and similarly for the sines (since the angles differ by precisely 2π) so that the root given by $-k$ will already have been accounted for by $n-k$. In practice, however, it is frequently found more convenient to consider $k = 0, -1, +1, -2, +2, -3, \ldots$, as at first above.

It is now a simple matter to deduce

THEOREM. *If p and q are integers, $q \neq 0$, then*

$$(\cos \theta + i \sin \theta)^{p/q} = (\cos p\theta + i \sin p\theta)^{1/q}$$
$$= \cos [(p\theta + 2k\pi)/q] + i \sin [(p\theta + 2k\pi)/q] \quad (k = 0, \pm 1, \pm 2, \ldots).$$

The theorem holds for both positive and negative rational indices. It should be pointed out that in this proof there is no loss of generality involved in taking q to be positive; since if p/q were negative we could take p as negative and q positive. In the case when p/q is a positive integer, the theorem reduces to De Moivre's theorem.

We are now in a position to be able to solve the equation $z^n = a$, where n is a positive integer and a any complex number. We obtain precisely n different roots in the field of complex numbers. In fact, if $a = r (\cos \phi + i \sin \phi)$ and $z = R (\cos \theta + i \sin \theta)$, we have the identity

$$R^n(\cos n\theta + i \sin n\theta) = r(\cos \phi + i \sin \phi),$$

whence

$$R^n = r, \quad n\theta = \phi + 2k\pi \quad (k = 0, \pm 1, \pm 2, \ldots),$$

giving, as above, precisely n different solutions

$$z = r^{1/n}[\cos \{(\phi + 2k\pi)/n\} + i \sin \{(\phi + 2k\pi)/n\}]$$

$$[k = 0, 1, 2, \ldots, (n-1)].$$

The equation $z^n = 1$ is of particular interest. The solutions of this equation are defined to be the nth roots of unity. These are $\cos 2k\pi/n + i \sin 2k\pi/n$ $[k = 0, 1, 2, \ldots, (n-1)]$. In the complex plane the points representing these numbers are seen to be equally spaced round the circumference of a circle of unit radius with centre the origin—the unit circle. If we write the first two roots as $1, \omega_n$, we see that we may write the set of roots as $1, \omega_n, \omega_n^2, \omega_n^3, \ldots, \omega_n^{n-1}$, whence all the roots are non-negative powers of ω_n. (Check this.)

More generally, the roots of $z^n = a$ (a complex) are represented by points equally spaced around the circumference of a circle with radius $|a|^{1/n}$ and centre 0, and the roots of $(z-b)^n = a$ by similarly arranged points on a circle with centre b and the same radius $|a|^{1/n}$.

In particular we can now, as stated in § 1.12, solve any quadratic equation $az^2 + bz + c = 0$ (a, b, c complex; $a \neq 0$): we first reduce it to the form $(z-d)^2 - e = 0$ and then apply the above with $n = 2$.

Exercises

1.1.1. Attempt to devise your own rules for addition and subtraction of p-numbers. In your system, are these results true?

(a) $(a+b)1 - b = a$.

(b) $a - a = 0$.

(c) $a + b = b + a$.

(d) $(a+b) + c = a + (b+c)$.

1.2.1. By considering both distance from 0 and direction from 0, try to find a method of ordering p-numbers. Examine your system for any unsatisfactory features. Do points which are "close" on the plane have nearly the same "size" according to your system, and conversely?

1.2.2. Let P represent a p-number, and drop a perpendicular from P to the real number line l to meet it at N. Consider as alternatives to OP the following definitions of "the modulus" of P: (a) ON; (b) PN; (c)$ON+PN$; (d) ON^2+PN^2; (e) 1, $(P \neq 0)$, 0, $(P = 0)$. Invent some others. In each case, draw the graph of the "circle" consisting of all points whose "modulus" is 1.

1.3.1. Using rectangular cartesian coordinates, plot the points: $(3, -3)$; $(3, 4)$; $(3, 0)$; $(0, 3)$; $(-3, 4)$; $(-3, 3)$; $(-3, 0)$; $(3, 3)$; $(-3, -3)$; $(-3, -4)$; $(0, -3)$. Taking them to represent p-numbers, and the x-axis as the real number line, calculate $\tan \theta$, where θ is the principal amplitude of the p-number in each case. Calculate the modulus also. In any case, where two distinct p-numbers have the same modulus and the same $\tan \theta$, show that either $\sin \theta$ or $\cos \theta$ differs for the two.

1.3.2. What can be said about the geometrical representation of p-numbers with (a) the same amplitude, (b) equal and opposite amplitudes, (c) amplitudes which differ by (i) π, (ii) 2π, (iii) 3π? What is the principal amplitude of a negative real number? Of a positive real number?

1.3.3. A regular hexagon has its centre at 0 and one of its vertices represents a p-number with modulus r and amplitude θ. Give the modulus and amplitude of the numbers represented by the other vertices.

1.4.1. (a) Calculate the modulus and principal amplitude of the following p-numbers: $(1 ,1)$; $(5, -5)$; $(-5, 12)$; $(-4, 0)$; $(-1, \sqrt{3})$; $(0, -4)$. (b) Determine the p-numbers whose modulus and amplitude are respectively (i) $5, \pi/2$; (ii) $\sqrt{2}, 3\pi/4$; (iii) $6, \pi$; (iv) $2\pi, 2\pi$; (v) $2\sqrt{3}, \pi/3$.

1.4.2. Invent some rules for multiplication of number pairs. Try to find any disadvantages each possesses. Can you divide also? Do you get the correct results for real numbers by your method if you consider $a \times b$ as $(a, 0) \times (b, 0)$? Do you obtain the result $(?, 0) \times (x, y) = (2x, 2y)$?

1.4.3. Do you consider that it would be possible to construct a sensible arithmetic of ordered number triples—(x, y, z)? Suggest some rules for addition and subtraction. (No harm in looking into multiplication, but it presents some difficulties.) Can you suggest a physical representation of triples and of these processes? Can you generalize further? If so, can you think of any practical applications?

1.5.1 Prove results (a), (b), and (c) in § 1.5 (p. 7).

1.5.2. If $(3, -7)+z = (2, 0)$, find the p-number z. Find z if the right-hand side of the equation is (a) $(-3, 7)$, (b) $(0, 0)$, (c) $z+z$, (d) $(3, -7)$.

1.5.3. Evaluate $(6, 7)+(-4, 2)+(-1, -8)$ in two different ways.

1.5.4. If $(a, b)+(p, q) = (0, 0)$, what is the relationship in the complex plane between the points (a, b) and (p, q)?

1.5.5. If $z = (2, 1)$ plot the points corresponding to $z, z+z, z+z+z$. Do the same for a few other values of z and suggest a general rule. What are the coordinates of w if $w+w+w+w = (2, 1)$?

1.6.1. Show geometrically that:

(a) If k is any positive integer and $z = (x, y)$, then $kz = (kx, ky)$.
(b) If p is a positive integer and $z = (x, y)$, then $(1/p)z = (x/p, y/p)$.
(c) If p and q are positive integers and $z = (x, y)$, then $(p/q)z = (px/q, py/q)$.

1.6.2. If z is a p-number and k is real and if $kz = (0, 0)$, what conclusions can be drawn?

1.7.1. If P represents z_1 and Q represents z_2 in the complex plane prove that mod (z_1-z_2) is given by the length of PQ and that amp (z_1-z_2) is given by the angle made by the directed line segment QP with the positive direction of the real axis.

Given amp $(z_1-z_2) = \theta$, find amp (z_2-z_1).

1.7.2. Solve the equations:

(a) $2z+(5, -7) = 3z-(-1, 11)$.

(b) $z+w = (8, 6),$
$\quad z-w = (2, -2).$ }

(c) $3z-4w = (6, -9),$
$\quad z+2w = (2, 7).$ }

Devise and solve some more linear equations.

1.7.3. If r_1 and r_2 are the moduli of z_1 and z_2 respectively, between what limits must mod (z_1-z_2) lie? Illustrate geometrically. Find two p-numbers z_1, z_2 such that mod $z_1 = $ mod $z_2 = $ mod $(z_1-z_2) = 2$. Is your solution unique?

1.8.1. Evaluate (a) $(5, 0)\times(0, 5)$; (b) $(-3, 5)\times(4, -2)$; (c) $(1, 1)\times(1, 1)$; (d) $(9, -2)\times(-9, -2)$; (e) $(2, 2)\times(2, 2)\times(2, 2)\times(2, 2)$; (f) $(5, 3)[(4, 2)+(-3, 7)]$; (g) $(4, 3)\times(-6, 1)\times(1, -2)$.

1.8.2. Solve the equation $(3, 2)z = (-4, -7)$. (Hint: let $z = (x, y)$ and expand the left-hand side; then compare with the right-hand side.)

1.8.3. If $(a, b)\times(x, y) = (1, 0)$, express x and y in terms of a, b.

1.8.4. By writing z_1 as (x_1, y_1), etc., prove directly that

(a) $z_1(z_2 z_3) = (z_1 z_2)z_3$.

(b) $\quad z_1 z_2 = z_2 z_1$.

(c) $z_1(z_2+z_3) = z_1 z_2 + z_1 z_3$.

(d) $(z_2+z_3)z_1 = z_2 z_1 + z_3 z_1$.

1.8.5. (a) If $z^2 = (0, 1)$ find all possible values for z.
(b) Evaluate $(-1/\sqrt{2}, -1/\sqrt{2})^2$.

1.8.6. Find a p-number (x, y) such that $(x, y)^2 = (5, 12)$. Is your solution unique? (x, y) is called a "square root" of (a, b) if $(x, y)^2 = (a, b)$. Consider the geometrical interpretation of a square root. Investigate cube roots, etc., from a geometrical point of view.

1.8.7. Find p-numbers $z = (x, y)$ satisfying the equation

$$z^2 - 3z + (2, 0) = (0, 0).$$

Verify your solutions.

Solve $z^2 + 2z + (2, 0) = (0, 0)$. [Hint: rewrite $\{z + (1, 0)\}^2 = (-1, 0).$]

Solve $z^2 - 2bz + c = (0, 0)$, where b and c are p-numbers.

1.9.1. If $(x_1, y_1) = (x, y) \times (x_2, y_2)$ and $(x_2, y_2) \neq (0, 0)$, prove, by expanding the right-hand side of the equation and solving two simultaneous equations, that

$$x = (x_1 x_2 + y_1 y_2)/(x_2^2 + y_2^2), \quad y = (y_1 x_2 - x_1 y_2)/(x_2^2 + y_2^2).$$

1.9.2. Evaluate the following: (a) $(1, -3) \div (0, -1)$; (b) $(5, 0) \div (4, 3)$; (c) $(10, -5) \div (-2, 1)$; (d) $(x^2 + y^2, 0) \div (x, y)$; $(1, 0) \div (x, y)$.
Invent and solve some more division problems involving p-numbers.

1.9.3. Verify that $(a, b) \times [(1, 0) \div (c, d)] = (a, b) \div (c, d)$. $[(c, d) \neq (0, 0).]$

1.10.1. Evaluate, giving your answer in the form $a + ib$, (a) $(4 + 3i)(4 - 3i)$; (b) $(7 + 2i)(3 + 10i)$; (c) $(2 - 5i)/(1 + i)$; (d) $(-\frac{1}{2} + \frac{1}{2}i\sqrt{3})^3$; (e) $(-i)^4$; (f) $(1 + i)/(1 - i)$.

1.10.2. Solve completely the equations

(a) $z^2 - 5iz - 6 = 0$; (b) $z^4 = 1$; (c) $z^3 - 1 = 0$. (Hint: factorize.)

1.11.1. Calculate the multiplicative inverses of (a) $1 + i$; (b) $4 - 3i$; (c) $4 - 2i$; (d) $4 + 2i$; (e) -1; (f) i; (g) $7 + 5i$.

1.11.2. Use 1.11.1 to simplify the expressions

(a) $[-i/(1 + i)]^2$; (b) $(3 + 2i)/(4 - 2i) + (3 - 2i)/(4 + 2i)$.

1.11.3. Determine the modulus and principal amplitude of the following complex numbers:

(a) -3; (b) $-i$; (c) $-1 + i$; (d) $\sqrt{2} - i\sqrt{2}$; (e) $-1 - i\sqrt{3}$.

11.1.4. Prove in detail that the complex numbers form a field under addition and multiplication.

1.12.1. Evaluate $i + i^2 + i^3 + i^4$ and illustrate the result geometrically.

1.12.2. Solve the equation $(1 + 2i)z^2 - 10iz + (-6 + 8i) = 0$. (See Exercise 1.9.3.) Give the general solution of $az^2 + bz + c = 0$, where a, b, and c are complex numbers $(a \neq 0)$.

1.12.3. Express $(1 + 2i + 3i^2 + 4i^3)/(4 - 3i + 2i^2 - i^3)$ in the form $a + ib$ (a, b real).

1.13.1. Prove the following results (i) algebraically, and (ii) geometrically:

(a) $z\bar{z} = x^2+y^2 = r^2$, where $z = x+iy$ and $r = |z|$.

(b) $\overline{z_1 \pm z_2} = \bar{z}_1 \pm \bar{z}_2$.

(c) $\overline{z_1 z_2} = \bar{z}_1 \bar{z}_2$, $\overline{(z_1/z_2)} = \bar{z}_1/\bar{z}_2$.

(d) $(z+\bar{z})/2 = \mathcal{R}(z)$, $(z-\bar{z})/2i = \mathcal{I}(z)$.

(e) $|\mathcal{R}(z)| \leqslant |z|$, $|\mathcal{I}(z)| \leqslant |z|$, equality occurring in the first case if z is real, and in the second case if z is imaginary.

1.13.2. Show that (a) $(\bar{z})^n = (\overline{z^n})$, where n is any non-negative integer; (b) if all the a_r, and z, are complex, then the conjugate of

$$\sum_{r=0}^{n} a_r z^r \quad \text{is} \quad \sum_{r=0}^{n} \bar{a}_r \bar{z}^r;$$

(c) if a polynomial equation with real coefficients has a complex root, then the complex conjugate of this root is also a root.

1.13.3. Prove that if z_1+z_2 and $z_1 z_2$ are both real, then z_1 and z_2 are either conjugate or both real. (Hint: consider a suitable quadratic equation.)

1.13.4. Identify the loci in the complex plane which are represented by the equations (a) $z = \bar{z}$; (b) $z+\bar{z} = 0$; (c) $|z| = r > 0$; (d) $|z-a| = b$ (a complex, b real and positive); (e) $|z-a| = |z-b|$ (a, b complex); (f) $az+\bar{a}\bar{z}+c = 0$ (a complex, c real); (g) $z\bar{z}+az+\bar{a}\bar{z}+c = 0$ (a complex, c real).

1.13.5. Express the equation $x^2/a^2+y^2/b^2 = 1$ (x, y, a, b real) in terms of $z, \bar{z}, a, b$. (Hint: $z+\bar{z} = 2x$.)

1.14.1. Show geometrically that $|z_1+z_2| \leqslant |z_1|+|z_2|$, equality occurring if z_1 and z_2 have the same amplitude but not otherwise unless one of them is zero.

Show also that $|z_1-z_2| \geqslant |z_1|-|z_2|$ and examine the conditions for equality to hold.

Also prove the conditions for equality algebraically in each case.

1.14.2. Prove that $|z_1+z_2|^2+|z_1-z_2|^2 = 2|z_1|^2+2|z_2|^2$ and interpret the result geometrically.

1.14.3. (a) Prove, by considering moduli or amplitudes, that the triangle whose vertices are the points z_1, z_2, and z_3 in the complex plane is equiangular if and only if

$$(z_1-z_2)/(z_2-z_3) = (z_2-z_3)/(z_3-z_1) = (z_3-z_1)/(z_1-z_2).$$

Show that these equations reduce to the single equation

$$z_1^2+z_2^2+z_3^2-z_2 z_3-z_3 z_1-z_1 z_2 = 0.$$

(b) Find necessary and sufficient conditions on z_1, z_2, z_3, z_4 for them to be represented in the complex plane by the vertices of a square.

1.14.4. Find a relation between z_1, z_2, and z_3 if the points representing them in the complex plane are collinear.

1.14.5. Generalize the triangle inequality to obtain an inequality applying to n complex numbers. Prove by induction or otherwise that your generalization is valid.

1.15.1. Solve the equation $z^3 + 1 = 0$ and show that the roots are represented in the complex plane by the vertices of an equilateral triangle.

1.15.2. Find the square roots of (a) i; (b) $-i$; (c) -4; (d) $1+i$.

1.15.3. Find the quadratic equation of the form $z^2 + bz + c = 0$ whose roots are (a) $1, i$; (b) $i, 2i$; (c) $1+2i, 1-2i$; (d) $\alpha + i\beta, \alpha - i\beta$.

What do you notice about b and c in cases (c), (d) in contrast with cases (a), (b)?

1.15.4. Solve $z^5 + 1 = 0$, and hence or otherwise factorize the expression $z^5 + 1$ into three factors, one linear and two quadratic, with real coefficients.

1.15.5. Factorize $x^{2n} + 1$ into n real quadratic factors.

1.15.6. Find all complex values of $(1 + i\sqrt{3})^{3/4}$.

1.15.7. Deduce from De Moivre's theorem that (a) $\cos 3\theta = 4\cos^3 \theta - 3\cos \theta$, (b) $\sin 3\theta = 3\sin \theta - 4\sin^3 \theta$.

1.15.8. Use De Moivre's theorem to express $\cos 5\theta$ and $\sin 5\theta$ in terms of powers of $\cos \theta$ and $\sin \theta$ respectively.

1.15.9. If $z = \cos \theta + i\sin \theta$, show that $1/z = \cos \theta - i\sin \theta$.

Deduce that $\cos \theta = \frac{1}{2}(z + 1/z)$, $\sin \theta = \frac{1}{2i}(z - 1/z)$,

$$4\cos^2 \theta = z^2 + 1/z^2 + 2 = 2\cos 2\theta + 2,$$

$$-8i\sin^3 \theta = (z^3 - 1/z^3) - 3(z - 1/z)$$

$$= 2i\sin 3\theta - 6i\sin \theta.$$

Express $\cos^6 \theta$ in terms of cosines of multiple angles. Obtain some other formulae of this type.

1.15.10. By means of multiplication by the factor $z - 1$ or otherwise, solve the equation

$$z^5 + z^4 + z^3 + z^2 + z + 1 = 0.$$

CHAPTER 2

POINT SETS IN THE COMPLEX PLANE.
SEQUENCES. LIMITS

2.1. Point Sets: Finite,
Countable, and Non-countable Sets.
Real Intervals

By a point set in the complex plane we merely mean a collection of points, for example:

(a) the (real) integers;
(b) the (real) integers of modulus less than or equal to 26;
(c) the points z satisfying $|z| = 5$;
(d) the points z satisfying $|z| < 5$;
(e) the real points z satisfying $|z| = 5$;
(f) the real points z satisfying $|z| < 5$;
(g) the seven points $\{44, i, -3, 2-6i, 29, -37i, 1\}$;
(h) the points $(2+3i)^n$, where n runs through all (real) integers;
(i) the points $(2+3i)^q$, where q runs through all rational numbers;
(j) the points satisfying $|z| = 5$ and $\mathcal{R}(z) = \pm 2$;
(k) the points z satisfying $z^{5/8} = a$ $(a \neq 0)$.

There are certain properties which immediately distinguish some of these sets from the others. First notice that each of the sets (b), (e), (g), (j), and (k) contains only a finite number of points. They contain $53, 2, 7, 4$, and 5 points respectively (check this). Each of the other sets contains an infinite number of points. A set containing a finite number of points is called a *finite set*. The sets (a), (c), (d), (f), (h), and (i) are thus not finite sets. However, let us consider (a); we can

30

set up a one-to-one correspondence between the natural numbers $1, 2, 3, 4, \ldots$, and the points of (a) by letting $1, 2, 3, 4, 5, 6, 7, \ldots$, correspond, for example, to $0, 1, -1, 2, -2, 3, -3, \ldots$, respectively, and in general letting the natural number N correspond to the integer $\frac{1}{2}N$ (N even) or $-\frac{1}{2}(N-1)$ (N odd). We have thus succeeded in "numbering off" the points in (a). A set which can be "numbered off" in this way is called "enumerable" or "countable". The set (h) above is thus also countable. Some results which are easily proved are the following:

(i) a set contained in a finite set is finite;

(ii) a set contained in a countable set is countable (or finite);

(iii) a set which can be divided up into a finite number of countable sets is countable;

(iv) a set which can be divided up into a countable number of finite sets is countable;

(v) a set which can be divided up into a countable number of countable sets is countable.

From (v) it follows that the set of all rational numbers is countable (see Exercise 2.1.1). With (iv) this then implies that the set (i) above is countable.

The remaining sets (c), (d), and (f) are, in fact, not countable (non-enumerable, or non-countable). We do not give a proof here. It depends on the fact that the set of all real numbers is non-countable; for the proof of this we refer to an elementary text on analysis such as reference 1, p. 108. (We may clearly not represent a non-countable set by a notation such as $\{z_1, z_2, \ldots\}$ which is applicable only to countable—including finite—sets.)

Among non-countable sets an important family is the family of *intervals* on the real axis. The set of all points representing real numbers x for which $a < x < b$, where a and b are given (real) constants with $a < b$, forms what is called the *open interval* $]a, b[$. [The alternative notation (a, b) often employed causes some slight danger of confusion with the complex number ($=$ real number pair) (a, b), although normally the context removes such danger.] If the end points

are included the interval is said to be *closed*, and is written in the form
[a, b]. Thus [a, b] is the set of points x such that $a \leqslant x \leqslant b$. We can
talk also of *half-open* (or *half-closed*) *intervals*]a, b] and [a, b[, representing the set of all x such that $a < x \leqslant b$, and $a \leqslant x < b$ respectively. By abuse of language we also use the notation]a, ∞[, ([a, ∞[)
to denote the set of x such that $x > a$, $(x \geqslant a)$, and]−∞, b[, (]− ∞, b])
for the set of x such that $x < b$, $(x \leqslant b)$. We call these sets (infinite)
intervals also.

If $a < b$ we shall find it convenient to say that a lies *to the left of b*
(or that b lies *to the right of a*) on the real axis.

2.2. Bounded and Unbounded Sets
on the Real Line

Let S be a set of points on the real line. If a number x_1 exists such
that $x \leqslant x_1$ for every x in S, the S is said to be *bounded above*, and x_1
is called an *upper bound* for S. If there are no numbers less than x_1
which are also upper bounds for S, then x_1 is called the *least upper
bound* of the set S (l.u.b. for short).

The *greatest lower bound* (g.l.b.) for a set which is bounded below
is defined similarly. A set of points on the real line for which both
upper and lower bounds exist is said to be *bounded*. Otherwise the set
is said to be *unbounded*.

We give some examples:

(a) The set of points $x = 1/n$, where n is a positive integer. This is a
non-finite but countable set with a l.u.b. 1, which is a member of the
set, and a g.l.b. 0, which is not. Evidently the set is not an interval,
since there are many x satisfying $0 < x < 1$ which are not included
in the set.

(b) The set of all rational numbers x such that $1 < x$. This set is
bounded below (its g.l.b. is 1, which is not a member of the set) but
not above, so it is not a bounded set. Again it is not an interval, since
there are many points representing numbers greater than 1 which are
missing from the set (e.g. $\sqrt{2}$).

(c) The set of all real x such that $x \leqslant 1$. This set is bounded above, but not below, so is again an unbounded set. In this case, since all points to the left of 1 are included, it is an (infinite) interval.

It is very important to note that a bounded set of points on the real line may have no l.u.b. (or g.l.b.) WHICH IS A MEMBER OF THE SET. For example, an open interval $]a, b[$ has upper and lower bounds b, a which are the l.u.b. and g.l.b. respectively, neither of which is a member of the set. In examples (a) and (b) above, again, g.l.b.s occur which are not members of the set. However, a half-open interval $]a, b]$ does contain its l.u.b. although not its g.l.b., and a closed interval $[a, b]$ contains both its l.u.b. and g.l.b. The example (c) above contains its l.u.b, although no lower bound exists.

Notice that we have not yet proved that in general when a set is bounded above (below) it necessarily has a least upper (greatest lower) bound. The method of proof depends on one's original definition of the real numbers. We shall not discuss this here but simply assume the result to be true. (This is really equivalent to the assumption that there are no "gaps" in the real numbers.)

2.3. The Bolzano–Weierstrass Property

Bounded sets containing an infinite number of points possess an important property which we shall now consider. First of all we shall prove a useful result about families of closed intervals.

PROPOSITION. *Let* $\{I_n\}$, $n = 1, 2, 3, \ldots$, *be a collection of closed intervals with the property that* I_{n+1} *is contained within* I_n. *(We call such a collection a nested family of intervals.) Then there is a point* c *which belongs to all the intervals* $I_1, I_2, \ldots$, *simultaneously.*

Proof. Let a_n, b_n $(a_n < b_n)$ be the end points of the interval I_n. Since I_{n+1} is contained in I_n we have

$$a_1 \leqslant a_2 \leqslant a_3 \ldots \quad \text{and} \quad b_1 \geqslant b_2 \geqslant \ldots.$$

Let c be the l.u.b. of the set A of points $\{a_1, a_2, a_3, \ldots\}$. Since c is an upper bound for A we have $c \geqslant a_n$ for each n. Now every b_n is an upper bound for A since the intervals are nested, and since c is by definition the l.u.b. for A we have $c \leqslant b_n$ for every n. Thus $a_n \leqslant c \leqslant b_n$ for every n, which is the same as saying that c belongs to I_n for every n. This proves the proposition.

Let S denote an infinite but bounded set of points on the real line. Let a be any lower bound and b be any upper bound, so that S is contained in $[a, b]$. If we divide the interval into two equal parts and consider the closed intervals $[a, (a+b)/2]$ and $[(a+b)/2, b]$ it is clear that an infinite number of points of S must lie in at least one of these intervals—possibly both. We now select the interval which contains an infinite number—either interval, if both will serve—and repeat the process of dividing into two and selection. By continuing this process we will, after n such divisions, arrive at an interval of width $(b-a)/2^n$ which still contains an infinite number of points of the set S. It is clear that if we choose any $\varepsilon > 0$, however small, we can, by sufficiently increasing n, ensure that $(b-a)/2^n < \varepsilon$. Now by the above proposition there is a point c which will lie within all the members of this family of nested intervals. Since, as above, these intervals are arbitrarily small and each contains by construction an infinite number of points of S, there are an infinite number of points of S within any arbitrarily small distance ε (> 0) of c.

A point such as c is called a *limit point*, or an *accumulation point*, of S; and this property that every bounded infinite set of real numbers has at least one limit point, is often called the *Bolzano–Weierstrass property* (BWP).

It should be noted that other limit points besides c may or may not exist, and if they do, may or may not be points belonging to S. For example, the set of rational numbers of form $1/n$ ($n = 1, 2, 3, \ldots$) is bounded and has precisely one limit point 0 which is not a member of the set; while the open interval $]0, 1[$ consists entirely of limit points which are members of the set. The end points of this interval are also limit points, but they are not members of the set of points in the interval.

2.4. Bounded and Unbounded Sets
in the Complex Plane

Since the complex numbers are not ordered, we need to proceed slightly differently in defining boundedness of sets of complex numbers. (We cannot define "upper" and "lower" bounds as for sets of real numbers.) If a circle with centre the origin and (finite) radius R exists, which contains in its interior all the points of a given set S, then S is said to be bounded. Evidently this amounts to the condition $|z| < R$ for all z in S.

The BWP still holds in the following sense.

THEOREM. *Given any bounded infinite set S of points in the complex plane, we can find at least one point in the plane (not necessarily a member of S) which contains an infinite number of points of S within any arbitrarily small distance ε (> 0) from it.*

Proof. Since S is bounded we can find real numbers a, b, c, d ($a < b$, $c < d$) such that S lies in the rectangle R_1 of points z such that $a \leqslant \mathcal{R}(z) \leqslant b$ and $c \leqslant \mathcal{J}(z) \leqslant d$. Divide the rectangle into four sub-rectangles of width $(b-a)/2$ and height $(d-c)/2$ by joining mid-points of opposite sides. One (or more) of these rectangles must contain an infinite number of points of S. Select one such rectangle (call it R_2) and subdivide again. Continuing this process we obtain a collection $R_1, R_2, R_3, \ldots$, of rectangles, each containing an infinite number of points of S, such that R_{n+1} is contained in R_n, and has width $(b-a)/2^{n+1}$ and height $(d-c)/2^{n+1}$. If R_n is the set of z with $a_n \leqslant \mathcal{R}(z) \leqslant b_n$ and $c_n \leqslant \mathcal{J}(z) \leqslant d_n$, then $\{[a_n, b_n]\}$ is a nested family of closed intervals, so that by the proposition there exists an f such that $a_n \leqslant f \leqslant b_n$ for all n; and $\{[c_n, d_n]\}$ is also a nested family so that there exists a g such that $c_n \leqslant g \leqslant d_n$ for all n. The point $z = f + ig$ therefore belongs to every rectangle R_n, and since the R_n become arbitrarily small as n increases it is clear that $f + ig$ is a limit point for S.

As in the real case, this method of constructing a limit point leaves undetermined whether there are or are not any other limit points.

2.5. Neighbourhoods. Open Sets

It is frequently necessary to refer to the points in the complex plane which, in some sense, are "near" a given point. To make this idea more precise, we define an ε-*neighbourhood* of a point z_0, where ε is any positive real number, to be the set of all points z such that $|z - z_0| < \varepsilon$. (The number ε is usually, but not invariably, considered as being "small".) All such points lie in the interior of a circle with centre z_0 and radius ε. We similarly define the (real) ε-neighbourhood of a point x_0 on the real axis, when considered as a real, not a complex number, to be the open interval $]x_0 - \varepsilon, x_0 + \varepsilon[$. (Fig. 2.1.)

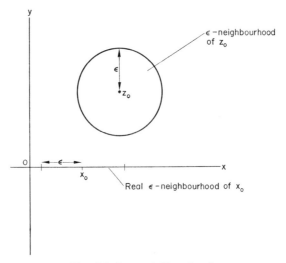

FIG. 2.1. An ε-neighbourhood.

DEFINITION. *If a set D in the complex plane is such that for every point s in S there is some $\varepsilon > 0$ such that the ε-neighbourhood of s consists entirely of points of S, then S is said to be an* open set *in the complex plane, and every such point s is said to be an* interior point.

Similarly, if the set T in the real line is such that every point t in T has an ε-neighbourhood consisting of points of T, then T is said to be an open set in the real line. Note that if we are dealing with the real line only, the "open intervals" are evidently open sets, whereas if we are dealing with all the points in the complex plane, then intervals on the real line are not open sets (in the complex plane). This is because neighbourhoods of points are now interiors of circles, containing points not on the real line.

In the complex plane, the interior of any polygon, or of any simple closed curve, is an open set. The empty set, consisting of no points whatsoever, is conventionally regarded as being open. From the definition, it is easy to see that the complex plane is an open set in itself, and that the real line is an open set in itself; but the real line is not an open set in the complex plane.

2.6. Limit Points

We have already (§ 2.3) met the idea of a limit point. We now use the idea of "neighbourhood" in order to rephrase the definition.

DEFINITION. *If for all $\varepsilon > 0$ the ε-neighbourhood of a point z_0 contains at least one point of a given set S, apart from (possibly) z_0 itself, then the point z_0 is said to be a* limit point, *or* accumulation point *of S.*

The point z_0 itself may or may not be a member of S. Consider, for example, on the real line the points corresponding to $x = 1 \pm 1/n$ ($n = 1, 2, \ldots$). The point $x = 1$ is clearly not a member of the set, yet is a limit point of it.

It is easy to see from the definition that in fact any ε-neighbourhood of a limit point of S must actually contain an infinite number of points of S. For if it contains a point z_1 ($z_1 \neq z_0$) and $|z_1 - z_0| = \varepsilon'$, consider the neighbourhood $|z - z_0| < \varepsilon'$. This must contain a point z_2 ($\neq z_0$) of S. Let $|z_2 - z_0| = \varepsilon''$ and consider the neighbourhood $|z - z_0| < \varepsilon''$; and so on. The process can evidently be repeated indefinitely, producing an unending collection of points $z_1, z_2, z_3, \ldots$, all lying within the original (arbitrary) ε-neighbourhood.

2.7. Closed Sets

If a set contains all its limit points it is said to be *closed*. The "closed intervals" referred to in § 2.1 are closed sets on the real line. An example of a closed set in the complex plane is the set of points z given by $|z-z_0| \leqslant r$. The ε-neighbourhood of a point Z is not closed, as points at a distance exactly ε from Z are limit points of the neighbourhood, but are not included in it. We regard the empty set as closed, since it (trivially) satisfies the criterion (there are no limit points, so it contains them all!). Evidently the set of all points in the complex plane is closed. These two examples show that sets may be both closed and open. (In fact they are the only two sets in the complex plane which are both closed and open.) Sets need not be either open or closed. See Exercises 2.5.1 and 2.7.1.

The following proposition shows the connection between the notions of open and closed sets.

PROPOSITION. *Let S_1 be any set in the complex plane (or real line) and let S_2[†] be the set of all those points in the complex plane (real line) which do not lie in S_1. Then S_1 is closed if and only if S_2 is open.*

Proof. Suppose S_1 is closed, and choose z_0 in S_2. Since z_0 is not in S_1 it cannot be a limit point of S_1 (for S_1 contains all its limit points), so some ε-neighbourhood of z_0 contains no points of S_1, i.e. lies entirely in S_2. This is true for all z_0 in S_2, so S_2 is by definition open. On the other hand, if S_1 is not closed then it has a limit point z_1 not belonging to it, i.e. belonging to S_2. Every ε-neighbourhood of z_1 contains points in S_1 (by definition of limit point) so no ε-neighbourhood of z_1 lies in S_2. Hence S_2 is not open.

2.8. Boundary Points

DEFINITION. *A point z_0 such that every neighbourhood of it contains at least one point (possibly z_0 itself) of a given set S and also at least one point not in S is called a* boundary point *of S.*

† S_2 is called the *complement* of S_1.

Observe that a boundary point of S either belongs to S or is a limit point of S. Therefore every closed set S contains its boundary points. On the other hand, every open set S contains none of its boundary points (since every ε-neighbourhood of every boundary point contains points not in S).

2.9. Closure

DEFINITION. *The closure of a set S, denoted by $\bar{S}$, is the set consisting of the set S together with all its limit points. If S is a closed set, then clearly $\bar{S} = S$. Conversely, if $\bar{S} = S$, then S is closed.*

It is easy to check that the closure of S consists precisely of S together with its boundary points.

2.10. Sequences

Suppose we are given some rule which associates with each natural number $1, 2, 3, \ldots$ a complex number. (In the language of Chapter 4 this is just a complex-valued function, whose domain is the set $\{1, 2, 3, \ldots\}$.) We will denote by z_n the complex number associated by the rule with n, and we do not discount the possibility that $z_n = z_m$ for two or more distinct values n, m. The ordered set $\{z_1, z_2, z_3, \ldots\}$ is called a *sequence*, and often denoted just by $\{z_n\}$. Here n denotes the ordinal number of the particular term z_n (z_3 being the third term, for example). Occasionally, however, a slightly different notation is employed: we denote the first terms in the sequence by z_0, and in this case z_n is the $(n+1)$th term of the sequence. The context will always make clear where necessary which notation is involved.

2.11. Convergence

DEFINITION. *Let $\{z_n\}$ be a sequence. If a complex number L exists with the property that, for every positive ε, a number N may be found so that $|z_n - L| < \varepsilon$ for all n greater than N, then the sequence $\{z_n\}$ is said to be convergent and to have the limit L.*

For such a sequence we write $z_n \to L$ (z_n *approaches L*) *as n tends to infinity*, or $\lim\limits_{n \to \infty} z_n = L$. (Fig. 2.2.)

Two points should be noted. First, the number N will in general clearly depend on the ε selected, so that it would be more proper to replace N in the definition by $N(\varepsilon)$ to indicate the dependence. Secondly we note that the definition implies that L is the only number with this property with respect to $\{z_n\}$. If there were two, say L and M,

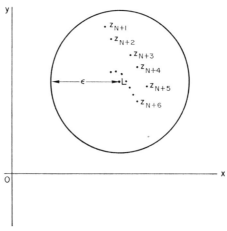

FIG. 2.2. Convergence of a sequence.

then choosing $\varepsilon = |L-M|/3$ we know that there exists N such that z_n is within ε of L if $n > N$, and also N' such that z_n is within ε of M if $n > N'$. Taking $m > \max (N, N')$ we see that z_m is within $|L-M|/3$ of both L and M, which is clearly impossible.

Let us now think of $\{z_n\}$ as just a set of points in the complex plane. We may ask what connection exists between this idea of a limit of the sequence $\{z_n\}$ and the idea which we have already met (§ 2.6) of a limit point of the set Z of points in the sequence. The (disappointing) situation is the following:

(i) $\{z_n\}$ may have a limit, even though Z has no limit point. Consider the sequence $\{1, 2, 3, 3, 3, 3, \ldots\}$. The set Z has no limit point since it consists only of the three points $\{1, 2, 3\}$. However the sequence clearly has 3 as a limit, since $|z_n - 3| = 0$ (which is less than any positive ε) for all $n > 2$.

(ii) Z may have a limit point, even although $\{z_n\}$ has no limit. Consider the sequence $\{1/2, 3/2, 1/4, 5/4, 1/8, 9/8, \ldots\}$ in which $z_{2n+1} = 1/(2^{n+1})$ ($n = 0, 1, 2, \ldots$), $z_{2n} = 1 + 1/2^n$ ($n = 1, 2, \ldots$). The set Z has two limit points, namely 0 and 1 (given any $\varepsilon > 0$ we can find some z_n such that $|z_n| < \varepsilon$, and some z_m such that $|z_m - 1| < \varepsilon$ —in fact an infinite number of each). However, the sequence has no limit since there are terms with n arbitrarily large such that $|z_n| < 1/3$, and also terms with n arbitrarily large such that $|z_n - 1| < 1/3$, so that $|z_n| > 2/3$.

2.12. Divergence

DEFINITION. *A sequence which is not convergent is said to be* divergent.

Divergent sequences may behave in one of three ways. In the first place a divergent sequence may be *bounded*, i.e. there may exist a real R such that $|z_n| < R$ for all n. If no such R exists, the sequence is said to be *unbounded*, and unbounded sequences may be of either of two types. It may be true that, given any R however large, there is some N such that $|z_n| > R$ for all n greater than N. (This N will depend on R.) Such a sequence is said to *diverge to infinity*. The sequence of natural numbers $\{1, 2, 3, 4, \ldots\}$ is an obvious example of this type of sequence. On the other hand, the sequence $\{1, 2, 1, 3, 1, 4, 1, \ldots\}$ does not satisfy the criterion of divergence to infinity, since for any R greater than 1 the point z_n will lie within the circle $|z| = R$ for all odd values of n, however large. This is the third type of divergence mentioned above, that in which the sequence is neither convergent nor divergent to infinity.

Note that if a sequence is convergent, with limit L, we say that it *converges* (to L). Similarly, a divergent series *diverges*.

2.13. Boundedness of Convergent Sequences

A useful although elementary result is the following.

THEOREM. *A convergent sequence is bounded.*

This is almost obvious, since the terms of a convergent sequence must eventually all be within ε of the limit L, so that they have no chance to "head off" to infinity. Here is an accurate proof.

Proof. Let the limit of the sequence be L; then if we take ε as 1, there exists a corresponding N such that $|z_n - L| < 1$ for $n > N$. Now let z_p be the term of greatest modulus among the first N terms of the sequence. (Any, if there is more than one.) Then the larger of the two concentric circles $|z| = |z_p| + 1$ and $|z| = |L| + 1$ contains all the points of the sequence. (It may be helpful to draw a diagram to clarify the argument.) There are two possibilities; either there is one of the first N terms whose modulus is as large as or larger than that of all the other terms, or there is not. In the latter case, it is the second circle which is required.

2.14. A Test for Convergence

It is almost self-evident that a sequence $\{z_n\}$ will converge if and only if both the sequences $\{\mathcal{R}(z_n)\}$ and $\{\mathcal{I}(z_n)\}$ converge. To show this formally, we note that if z_n and z_0 are complex numbers of form $x_n + iy_n$ and $x_0 + iy_0$ respectively, then

$$|z_n - z_0| = |x_n + iy_n - x_0 - iy_0| \leqslant |x_n - x_0| + |y_n - y_0|. \quad \text{(Why?)}$$

Hence if $|x_n - x_0| < \varepsilon/2$ and $|y_n - y_0| < \varepsilon/2$, then $|z - z_0| < \varepsilon$.

Thus if the real parts converge to x_0 and the imaginary parts to y_0 it follows that z_n converges to $x_0 + iy_0$.

On the other hand, since

$$|z_n - z_0|^2 = |x_n - x_0|^2 + |y_n - y_0|^2$$
$$\geqslant |x_n - x_0|^2 \quad \text{and} \quad \geqslant |y_n - y_0|^2,$$

it follows that if $|z_n - z_0| < \varepsilon$, then $|x_n - x_0| < \varepsilon$ and $|y_n - y_0| < \varepsilon$.

Now if z_n approaches z_0, then given $\varepsilon > 0$ we can find N such that $|z_n - z_0| < \varepsilon$, and hence $|x_n - x_0| < \varepsilon$, $|y_n - y_0| < \varepsilon$, for all $n > N$. It follows that x_n will approach x_0 and y_n will approach y_0. This proves the result.

2.15. Cauchy Sequences of Real Numbers

If a sequence $\{z_n\}$ is convergent (say $z_n \to a$), then since as n increases all the terms become "close to" a, they become "close to" each other. To put this precisely, given ε' we have $|z_n - a| < \varepsilon'$ if n is greater than some $N(\varepsilon')$, whence, if m, p are both greater than $N(\varepsilon')$, we have

$$\begin{aligned} |z_m - z_p| &= |(z_m - a) + (a - z_p)| \\ &\leqslant |z_m - a| + |a - z_p| \\ &\leqslant \varepsilon' + \varepsilon' = 2\varepsilon'. \end{aligned}$$

So, first choosing ε and then letting $\varepsilon' = \frac{1}{2}\varepsilon$, we have

$$|z_m - z_p| < \varepsilon \quad \text{if} \quad m, p > N.$$

It is natural to wonder whether, as seems likely, this provides a sufficient test for convergence. To be explicit, a sequence which possesses the property that for any $\varepsilon > 0$, there exists $N(\varepsilon)$ such that for all $n, p > N(\varepsilon)$,

$$|z_n - z_p| < \varepsilon$$

is called a Cauchy sequence; we would like to be able to state that a Cauchy sequence is convergent. This is in fact the case. We prove it first for sequences of real numbers.

THEOREM. *A Cauchy sequence of real numbers is convergent.*

Proof. We start by selecting some fixed positive δ and an N such that $|x_n - x_p| < \delta$ for all $n, p > N$ (the sequence being denoted by $\{x_n\}$). It follows that the sequence $\{x_{N+1}, x_{N+2}, x_{N+3}, \ldots\}$ is an infinite sequence all of whose terms lie in the closed interval $I = [x_{N+1} - \delta, x_{N+1} + \delta]$. Let S denote the set of real numbers which appear as terms in this sequence. There are two possibilities: (i) S is finite. In this case there must exist some number y in S such that $x_n = y$ for an infinite number of values of n—hence for values greater than any given N. Setting p equal to one of these in the definition of Cauchy sequence gives $|x_n - y| < \varepsilon$ for all $n > N$, which means that $\{x_n\}$ converges to y.

(ii) S is infinite. By the Bolzano–Weierstrass property the set S has a limit point P in the interval I. Because the sequence is a Cauchy sequence, given any $\varepsilon > 0$, we can choose M such that for every $n, p > M$, $|x_n - x_p| < \varepsilon/2$. Moreover, since there is a limit point P there is a value q of the suffix such that $q > M$ and $|x_q - P| < \varepsilon/2$. Then, for $n > M$

$$|x_n - x_p| + |x_q - P| < \varepsilon/2 + \varepsilon/2 = \varepsilon.$$

But

$$|x_n - P| \leqslant |x_n - x_q| + |x_q - P|,$$

and so, finally,

$$|x_n - P| < \varepsilon \text{ for all } n > M.$$

Since ε is arbitrary, this proves that $\{x_n\}$ converges to P.

2.16. Cauchy Sequences of Complex Numbers

A similar proof to the above applies in the case of complex numbers. For a given δ there is a circle $|z - z_n| = \delta$ (denoted by C) within which the z_n lie for all $n > N$. We let Z denote the set of points in the plane corresponding to the sequence $\{z_{N+1}, z_{N+2}, \ldots\}$. As before, there are two cases: (i) Z finite, (ii) Z infinite. In case (i) there is some complex number w such that $z_n = w$ for an infinite number of values of n, and the argument proceeds as in case (i) of § 2.15. In case (ii) we know by the BWP that within, or on, the circle C there is a limit point P, and the rest of the proof is word for word the same as in case (ii) of § 1.15, with z in place of x.

2.17. Non-decreasing Real Sequences

We conclude this chapter with an easily proved but very useful result on real sequences. Note that for real sequences we can define upper bound, l.u.b., etc., just as for sets of points on the real line.

THEOREM. *Any non-decreasing sequence of real numbers with an upper bound is convergent.*

Proof. Let the sequence be $\{u_n\}$ and let its l.u.b. be u. Then for every $\varepsilon > 0$ an N exists such that $u - \varepsilon < u_N \leqslant u$. (If this were not the case, i.e. if no u_N could be found greater than $u - \varepsilon$, then $u - \varepsilon$ would be an upper bound for the sequence, contrary to the hypothesis that u is the l.u.b.) Since the sequence is non-decreasing, it follows that $u - \varepsilon < u_n \leqslant u$ for all $n \geqslant N$. Rearranging, we see that $|u - u_n| < \varepsilon$ for all $n \geqslant N$, so the condition for convergence is satisfied, and the theorem is proved.

Exercises

2.1.1. Prove the statements (i) to (iv) in § 2.1.

For each $r = 1, 2, 3, \ldots$, let $S^{(r)}$ be a countable set, with its points "numbered off" $z_1^{(r)}, z_2^{(r)}, z_3^{(r)}, \ldots$ By considering the list

$$z_1^{(1)}$$
$$z_1^{(2)}, z_2^{(1)}$$
$$z_1^{(3)}, z_2^{(2)}, z_3^{(1)}$$
$$z_1^{(4)}, z_2^{(3)}, z_3^{(2)}, z_4^{(1)}$$
$$\cdot \quad \cdot \quad \cdot \quad \cdot \quad \cdot$$
$$\cdot$$
$$\cdot$$

or otherwise, prove the statement (v).

Deduce that the set of rational numbers is countable.

2.1.2. Considering the rationals m/n, km/kn as different (e.g. taking $1/2$ as distinct from $2/4$, etc.), devise an explicit method of enumeration for the rationals between 0 and 1 inclusive. (Hint: consider the sum of numerator and denominator.) Why cannot the rationals be enumerated in order of magnitude?

2.1.3. Decide which of the following intervals are (a) open, (b) closed, (c) half-open:

(i) All positive real numbers less than 1.
(ii) All non-negative real numbers not greater than 1.
(iii) All real numbers with a modulus less than 1.
(iv) All negative real numbers not less than -2.

By considering rectangles with edges parallel to the real and imaginary axes, devise a definition for "an open interval in the complex plane". Can you generalize further?

2.1.4. Let I and J be two intervals. Prove that the set of points (if any) lying in both I and J consists of either (a) one point or (b) an interval. Show that if I is open then (a) cannot arise. Investigate what possibilities can arise by taking I and/or J to be closed, half-open, etc.

2.2.1. Determine, if they exist, the least upper bounds and greatest lower bounds of the following sets of points on the real line, stating whether or not the bounds are themselves members of the set.

(a) All numbers of form $(n-1)/n$ where n is a non-zero integer.
(b) All numbers with modulus less than 100.
(c) All integers less than 100.
(d) All positive integers less than or equal to 10.
(e) All positive rational numbers with numerator less than denominator.
(f) All irrational numbers not less than 2.
(g) The intervals given in Exercise 2.1.3.

2.2.2. Examine whether each of the sets in Exercise 2.2.1 contains any limit point or points. State in each case whether any limit point you discover is a member of the set.

2.3.1. Give examples of sets of numbers between 0 and 1 which satisfy the following conditions, respectively:

(a) The number 1 is neither the l.u.b. nor a limit point.
(b) The number 1 is the l.u.b. and a limit point but not a member of the set.
(c) The number 1 is a member of the set but not a limit point.

Would it be possible to devise a set for which the number 1 was the l.u.b. yet neither a member of the set nor a limit point? Justify your answer.

Devise some similar problems relating to lower bounds, and try to solve them.

2.4.1. Which of the following sets in the complex plane are bounded?

(a) All numbers of form $i(n-1)/n$ (n a non-zero integer).
(b) All numbers of form i^n/n (n a non-zero integer).
(c) All numbers with amplitude $\pi/10$.
(d) All numbers with modulus $1/10$.
(e) All real integers less than 100.
(f) All numbers with real part less than 3.
(g) All numbers with real part positive and less than or equal to 3.
(h) All numbers as in (g) with imaginary part equal to ± 2.
(i) All numbers as in (g) with imaginary part rational.

2.4.2. Prove (a) a bounded set on the real line is also bounded when considered as a set in the complex plane, (b) if a set S is bounded in the complex plane, and T is contained in S, then T is bounded.

2.5.1. Determine which of these sets of complex numbers are open (in the complex plane):

(a) All real numbers between 0 and 1 inclusive.
(b) All real numbers between 0 and 1 exclusive.
(c) All points inside the circle $|z-1| = 3$.
(d) All complex numbers with moduli less than or equal to 100.
(e) All complex numbers with rational real parts.
(f) All complex numbers.
(g) All z such that $|z| = 1$.

2.6.1. Find any limit points of the following sets.

(a) All complex numbers of form $(1-1/n)+i(1+1/n)$, where n is a non-zero integer.
(b) The points z such that amp $z = \pi/4$.
(c) The points $1, i, \frac{1}{2}, i/2, \frac{1}{3}, i/3, \ldots, 1/n, i/n, \ldots$
(d) The points (r, θ) in polar notation, where $r = 1+1/n$, $\theta = \pi/n$ ($n = 1, 2, 3, \ldots$)

2.7.1. Which of the sets of Exercises 2.5.1 and of 2.4.1 are closed in the complex plane?

2.7.2. Show that any set containing only a finite number of points is both bounded and closed (a) in the complex plane, (b) (where the points are real) in the real line.

2.8.1. Identify any boundary points for the sets in Exercises 2.5.1 and 2.4.1.

2.8.2. Let S_1 be a set in the complex plane and let S_2 be the set of all points not in S_1. Show that every boundary point of S_1 is a boundary point of S_2, and vice versa.

2.8.3. Show that the set of boundary points of any set is closed.

2.8.4. Show that every point of a finite set S is a boundary point of S.

2.8.5. Show that if a set contains all its boundary points, then it must be a closed set. Deduce that if a set contains none of its boundary points then it must be an open set.

2.8.6. Let S be a set of points in the complex plane. If z_0 is a point in S and z_1 is a point not in S, prove that every path joining z_0 to z_1 contains at least one boundary point of S. (Hint: let such a path C be given by $z = z(t), 0 \leqslant t \leqslant 1$, with $z(0) = u_0$ and $z(1) = z_1$. Let T be the least upper bound of the set of those numbers t in the interval [0, 1] for which the point $z(t)$ on C is an element of S. Now show that $z(T)$ is a boundary point of S.)

2.9.1. Prove that the closure of a set is a closed set.

2.9.2. In Exercise 2.8.2 show that the boundary points of S_1 (or S_2) are precisely those points lying in $\bar{S}_1$ and $\bar{S}_2$.

2.9.3. Give an example of two sets A, B which are open and have no common point but for which $\bar{A}$ and $\bar{B}$ do have a common point. Give a second example in which neither A nor B is an open set.

2.10.1. On the real line, plot the first ten terms of each of the following sequences $\{z_n\}$.

(a) $z_n = 2+(-1)^n(1/n)$ $(n = 1, 2, \ldots)$.
(b) $z_n = 2-1/n$.
(c) $z_n = 2+1/n$.
(d) $z_{2n} = 2$, $z_{2n-1} = 2+1/n$.

Determine in each case how many terms have to be taken before all subsequent terms lie in the range $1.999 < z_n < 2.001$.

2.11.1. Let $\{z_n\}$ be a bounded sequence, no two of whose terms are the same. Using the BWP or otherwise, prove that if the set of points representing the terms of the sequence has only one limit point, then the sequence is convergent. Generalize to sequences none of whose terms is repeated more than a finite number of times.

2.11.2. Find examples of bounded sequences with 2, 3, 4 and an infinite number, respectively, of limit points.

2.11.3. Prove that if $\{u_n\}$ is a convergent sequence, and $\{v_n\}$ is a sequence such that $u_n = v_n$ except for a finite number of values of n, then $\{v_n\}$ is convergent.

2.11.4. Decide which of the following sequences are convergent; for those that are, state the limit L of the sequence. Taking $\varepsilon = 0.1$, find for each convergent sequence an N (which need not be the smallest possible) such that $|z_n-L| < \varepsilon$ for $n > N$.

(a) $z_n = 1+i/n^2$ $(n = 1, 2, \ldots)$.
(b) $z_n = n+i/n^2$.
(c) $z_n = (i/n)^n$.
(d) $z_n = (2-1/n)+i(\sin \pi/n)$.
(e) $z_n = 1/n+in\pi/(n+1)$.
(f) $z_n = \sin n\pi+i \cos n\pi$.

2.11.5. If $\{u_n\}$ and $\{v_n\}$ are convergent sequences, with limits U, V respectively, show that the sequences (i) $\{u_n+v_n\}$, (ii) $\{u_n-v_n\}$, (iii) $\{u_nv_n\}$ are also convergent and have limits $U+V$, $U-V$, and UV respectively. (Hint: let $u_n = U+\phi_n$. Show that ϕ_n tends to zero as n tends to infinity. Let $v_n = V+\theta_n$ similarly. Then $\phi_n, \theta_n < \varepsilon$ for $n > N(\varepsilon)$).

2.12.1. Devise examples of sequences $\{z_n\}$ which are (a) bounded but not convergent, (b) unbounded yet not diverging to infinity, (c) unbounded and containing no real terms yet diverging to infinity, although $\{\mathcal{J}(z_n)\}$ is convergent to zero, (d) divergent to infinity but in no particular direction (i.e. {principal amplitude of z_n} is not convergent). For example, $1, 2+2i, 3i, -4+4i, -5, -6-6i, -7i, 8+8i, 9, 10+10i, \ldots$.

2.12.2. Investigate the truth or falsity of the following assertion: "An unbounded sequence which does not diverge to infinity has at least one limit point."

2.12.3. Prove that, whether an infinite sequence $\{z_n\}$ of points in the complex plane is bounded or not, there is an angle α (not necessarily unique) such that, for any $\varepsilon > 0$, however small, an infinite number of points z_n have amplitudes θ_n such that $\alpha - \varepsilon < \theta_n < \alpha + \varepsilon$. (Hint: consider radial subdivision of the complex plane.)

2.13.1. Find the radius of the smallest circle bounding each of the following sequences $\{z_n\}$ of points in the complex plane:

(a) $z_n = (n^3+1)/n!$; (b) $z_n = (3i)^n/n!$; (c) $z_n = (1+1/n)+i/n^2$; (d) $z_n = (n-1)/n](1+i)$; (e) $z_n = 3-4/n+i(4-5/n)$.

2.14.1. Determine whether each of the following sequences $\{z_n\}$ converges, and, if so, find the limit in each case.

(a) $z_n = (n+1)/(n+2)+i(n+3)/(n+4)$,

(b) $z_n = \cos n\pi/4+i \sin n\pi/4$.

(c) $z_n = (1/n) \sin n\pi/4+i \cos n\pi/4$.

(d) $z_n = (1-1/n) \sin n\pi/4+i(1+1/n^2) \cos n\pi/4$.

(e) $z_n = (1/n) \sin n\pi/4+i[n^2/(n^3+1)] \cos n\pi/2$.

2.15.1. All points of a given set S in the complex plane are within a distance d from all other points in the set. Prove that the set is bounded and give a number R such that $|z| < R$ for all z in S. Determine also the four vertices of a square which would contain all points of the set. (Note: it will be necessary to bring in, e.g., $|z_0|$, where z_0 is some point in the set, as well as d, in determining R and completing the exercise.)

2.15.2. Let $\{z_n\}$, $\{w_n\}$ be two convergent sequences. Suppose that given $\varepsilon > 0$ there exists $N(\varepsilon)$ such that $|z_p - w_n| < \varepsilon$ whenever $p, n > N$. Deduce that $\{z_n\}$ and $\{w_n\}$ have the same limit.

2.15.3. Show that the sequence $\{s_n\}$, where $s_n = (1-r^n)/(1-r)$, is a Cauchy sequence provided that r lies within a certain range: hence deduce that the sequence converges for $|r| < 1$, and diverges otherwise.

2.16.1. By considering real and imaginary parts separately, prove that every Cauchy sequence of complex numbers is convergent.

2.16.2. Prove that the sequence $\{z_n\}$, where $z_n = \cos \pi/n+i \sin \pi/n$, is a Cauchy sequence. State the limit of the sequence.

2.17.1. Prove that a non-increasing sequence of real numbers which is bounded below is convergent, and that the limit of the sequence is its g.l.b.

2.17.2. Determine the limit of each of the following sequences, assuming subsequent terms to follow the pattern indicated by the given terms:

(a) $\{1.1, 1.11, 1.111, 1.1111, \ldots\}$

(b) $\{4, 3\frac{1}{2}, 3\frac{1}{3}, 3\frac{1}{4}, 3\frac{1}{5}, \ldots\}$

(c) $\{4, 3\frac{1}{2}, 3\frac{1}{4}, 3\frac{1}{8}, 3\frac{1}{16}, \ldots\}$

(d) $\{\frac{1}{3}, \frac{1}{5}, \frac{1}{7}, \ldots, 1/(2n+1), \ldots\}$

2.17.3. If $\{u_n\}$ is non-decreasing, and with an upper bound, and $\{v_n\}$ non-increasing, with a lower bound, are either of these properties shared by $\{u_n+v_n\}$? (u_n-v_n)? (u_nv_n)? (u_n/v_n)? (In the last case, it may be assumed that v_n is never zero.) Consider the convergence or otherwise of these sequences.

CHAPTER 3

INFINITE SERIES. TESTS FOR CONVERGENCE

3.1. The Sum of an Infinite Series

If we have a sequence $\{u_n\}$ of real or complex numbers, we may form from it a second sequence $\{s_n\}$ in which the terms are defined as follows:

$$s_1 = u_1, \quad s_2 = u_1 + u_2, \quad s_3 = u_1 + u_2 + u_3, \ldots,$$

$$s_n = u_1 + u_2 + \ldots + u_n = \sum_{r=1}^{n} u_r.$$

While this repeated addition defines s_n for any value of n, it is clear that the sum "to infinity" of the u_r would in this sense be meaningless, since it would involve a never-ending process of addition.

It may nevertheless happen that the sequence $\{s_n\}$ converges to a limit, say s. In this case we define the sum to infinity of the series $u_1 + u_2 + \ldots$ to be the number s, and say that the series *converges to s*. To restate this more formally:

DEFINITION. *The sum* $\sum_{r=1}^{n} u$ *is called the nth partial sum of the (infinite) series* $u_1 + u_2 + \ldots + u_r + \ldots$ *If the sequence* $\{s_n\}$ *of partial sums converges to s, then the series (denoted by* $\sum_{1}^{\infty} u_r$*) is said to converge, or to* be convergent, *and s is said to be its* sum to infinity.

A series for which no such s exists is said *to diverge* or *to be divergent*. In this case the partial sums $u_1, u_1 + u_2, \ldots$, may still be formed, but they do not "home" on to any particular number.

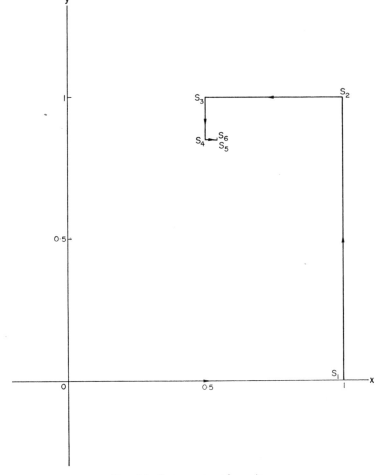

FIG. 3.1. Convergence of a series.

As an example of a convergent series (Fig. 3.1) let us consider the series

$$S = 1 + i + i^2/2! + i^3/3! + \ldots + i^n/n! + \ldots$$
$$= 1 + i - 1/2! - i/3! + \ldots.$$

It seems clear from the diagram that the partial sums become closer and closer to a particular point. This is the case, and, as will be found later, the point is approximately $0.54 + 0.84i$. Although the limit point is never reached, the discrepancy between it and the partial sum becomes increasingly negligible as we proceed.

On the other hand, if we consider the series

$$S = 1 + i + i^2 + i^3 + \ldots + i^n + \ldots$$
$$= 1 + i - 1 - i + \ldots$$

it is clear that however far the summation is extended there is no tendency to approach a particular number—more precisely, the sequence $\{s_n\}$ does not converge. Instead, we have cyclic repetitions of s_1, s_2, s_3, and s_4 ($1, 1+i, i, 0$ respectively).

Several different possibilities arise for the behaviour of the sequence of partial sums of an infinite series. As examples, we may find that:

(a) it converges towards some specific number (which we may or may not be able to calculate precisely—but if so it will be by some other method than never-ending addition!);

(b) the partial sums repeat a finite, or even an infinite, number of different values;

(c) the partial sums approach, but do not necessarily attain, several different values in turn (e.g. one limit for the sum of an odd number of terms and another for an even number);

(d) the partial sums form a bounded set ($|s_n| < R$ for some R; $n = 1, 2, \ldots$);

(e) the partial sums form an unbounded set, the members of which may either tend to lie near a particular line or lines, say, or may be more randomly distributed.

3.2. Summability

It should be noted that this method of "summing" (by convergence) an infinite series which, of course, cannot be summed in the normal sense of addition, is somewhat arbitrary. Other methods of "summing" also exist; methods which may be applied not only to convergent series but even to some which do not converge. A limiting process is still involved, but something other than the limit of $\{s_n\}$ is considered.

For example, let $S = 1-1+1-1+1-\ldots$. This has no sum by convergence, since s_n is alternately 0 or 1 as n is even or odd. Consider, however, the series

$$s = 1-x+x^2-x^3+\ldots$$

If $|x| < 1$, this can be "summed to infinity" by convergence, since $s_n = [1-(-x)^n]/(1+x)$, and this approaches the limit $1/(1+x)$ as n approaches infinity. If we now allow x to approach 1 "from below", the expression $1/(1+x)$ approaches $\frac{1}{2}$, while the expression $1-x+x^2-\ldots$ approaches $1-1+1-\ldots$ (in the sense that each term may be made to approach as close to 1 as we please by sufficiently increasing x without allowing it to take the value 1). We could thus adopt $\frac{1}{2}$ as, in some sense, the "sum" of the non-convergent series $1-1+1-\ldots$.

To objections that this is not what is meant by summing the actual series given, since by adding alternately $+1$ and -1 we never reach $\frac{1}{2}$, we would reply that summing by convergence is open to precisely the same objection; in the same way that we can never arrive at $\frac{1}{2}$ by repeated addition, we can never, for example, arrive at 2 by repeated addition of the terms

$$1+\tfrac{1}{2}+\tfrac{1}{4}+\tfrac{1}{8}+\ldots+1/2^n+\ldots.$$

Of course, in the latter case the fact that we know that by proceeding sufficiently far with our summation, $|2-s_n|$ may be made as small as we please—more precisely, that for any $\varepsilon > 0$, there is an $N(\varepsilon)$ such that $|2-s_n| < \varepsilon$ for all n greater than $N(\varepsilon)$—makes convergence seem a quite satisfactory method of summation. Also, for many purposes the "sum by convergence" is found to be more useful than a "sum" defined by any other method. We shall not in this book consider other methods of summation of infinite series, having merely drawn attention to their existence in order to emphasize that our definition of the sum of an infinite series is, like any other, a convention. We refer, for example, to G. H. Hardy (reference 3) for further information about this topic.

3.3. Testing for Convergence or Divergence

THEOREM. *Unless u_n approaches zero as n approaches infinity, the series $\sum_1^\infty u_n$ will not converge.*

Proof. If the series is convergent, then for any $\varepsilon > 0$ we can find an $N(\varepsilon)$ such that $|s_n - s| < \varepsilon/2$ for $n > N(\varepsilon)$, where s_n denotes the sum of the first n terms and s is the limit of $\{s_n\}$. It follows that for such values of n,

$$|s_{n+1} - s| < \varepsilon/2, \quad \text{whence} \quad |s_{n+1} - s_n| \leqslant |s_{n+1} - s| + |s - s_n|$$
$$\leqslant \varepsilon/2 + \varepsilon/2 = \varepsilon.$$

But $s_{n+1} - s_n = u_{n+1}$, hence $|u_{n+1}| < \varepsilon$, which is arbitrarily small for $n > N(\varepsilon)$; i.e. $|u_n| < \varepsilon$ for $n > N(\varepsilon) + 1$. Thus u_n approaches zero as n approaches infinity, as was to be proved.

It would be highly convenient if the tending of u_n to zero were a sufficient test for convergence of the sequence $\{s_n\}$; but this, unhappily, is not the case. The counter-example usually given is

$$1 + \tfrac{1}{2} + \tfrac{1}{3} + \ldots + 1/n + \ldots .$$

Here $u_n \ (= 1/n)$ approaches zero as n tends to infinity. However, the series does not converge, since if it had a sum S then evidently

$$S > 1 + \tfrac{1}{2} + \left(\tfrac{1}{4} + \tfrac{1}{4}\right) + \left(\tfrac{1}{8} + \tfrac{1}{8} + \tfrac{1}{8} + \tfrac{1}{8}\right) + \left(\tfrac{1}{16} + \ldots\right)$$
$$> \tfrac{1}{2} + \tfrac{1}{2} + \tfrac{1}{2} + \tfrac{1}{2} + \ldots ,$$

which clearly increases without limit, therefore giving a contradiction. Thus the property $u_n \to 0$ must on no account be used as a test indicating the convergence of $\sum_1^\infty u_n$.

(It is the author's experience that some students find it extremely difficult to persuade themselves of this in spite of the familiarity of the counter-example above. This may be due to some confusion between the series, which involves the addition of terms, and the sequence of individual terms; the sequence $\{u_n\}$ is, of course, convergent with limit zero—but this is not what is being considered. It is the effect of adding terms together which matters—not the diminution in size of individual terms; and a large number of exceedingly small terms has by no means necessarily a negligible total).

Obviously, if we can actually find a formula for s_n and show that s_n approaches a limit s as n approaches infinity, we are able to demonstrate convergence in the most convincing way. But, except in the case of a few simple series, formulae for the partial sums to n terms are usually not obtainable, and we need to have recourse to other tests. These will be discussed in subsequent sections, in which, in order to simplify notation, we shall write, instead of $\sum_{1}^{\infty} u_n$ for example, simply $\sum u_n$.

3.4. The Comparison Test

For series of real non-negative terms, probably the most important test for convergence is the comparison test. The essence of this test is to compare individual terms of the given series $\sum u_n$ with those of a series $\sum v_n$ which is already known to be convergent or divergent. The test may be stated in several forms, not entirely equivalent.

THEOREM A. *If $\sum v_n$ is a convergent series of non-negative terms, and if, for $n > N$, $0 \leqslant u_n \leqslant kv$, where k is some positive constant independent of n, then $\sum u_n$ is convergent.*

Proof. Let s_n denote $\sum_{1}^{n} u_r$ and S_n denote $\sum_{1}^{n} v_r$.

Since $\{S_n\}$ is convergent, it is a Cauchy sequence, i.e. $(S_{n+p} - S_n) < \varepsilon$ for $n > M(\varepsilon)$, $p \geqslant 0$. Select N_1 greater than the larger of M and N. Then

$$u_{n+1} + u_{n+2} + \ldots + u_{n+p} \leqslant k(v_{n+1} + v_{n+2} + \ldots + v_{n+p}) \quad \text{for} \quad n \geqslant N_1$$
$$\leqslant k(S_{n+p} - S_n)$$
$$< k\varepsilon.$$

If we now write $k\varepsilon = \varepsilon'$ this may be written in the form $(s_{n+p} - s_n) < \varepsilon'$ for $n \geqslant N_1$. Since ε' can be chosen arbitrarily small by suitably choosing ε, this establishes that $\{s_n\}$ is a Cauchy sequence, and therefore convergent. Thus $\sum u_n$ is a convergent series.

The corresponding theorem for divergence is set as an exercise (3.4.1).

THEOREM B. *If two series of positive terms, $\sum u_n$ and $\sum v_n$, exist such that u_n/v_n tends to a (finite) limit L, $(L > 0)$, then $\sum u_n$ and $\sum v_n$ are either both convergent or both divergent.*

Proof. It is easily seen, by taking $\varepsilon = \frac{1}{2}L$, that we may always choose a sufficiently large N so that, for $n > N$,

$$\tfrac{1}{2}L < u_n/v_n < 3L/2.$$

Then $\frac{1}{2}Lv_u < u_n < 3Lv_n/2$ for $n > N$, and the result follows as in Theorem A, by considering $s_{n+p} - s_n$.

With certain obvious modifications, this theorem can be extended slightly to take account of the cases when (i) $L = 0$, (ii) $u_n/v_n \to \infty$ with n. For example, if $\lim_{n \to \infty} u_n/v_n = 0$ and $\sum v_n$ is convergent, it is easy to show that $\sum u_n$ is convergent; while if the ratio tends to infinity with n and $\sum v_n$ is divergent, then $\sum u_n$ will also be divergent. These results are left as an exercise (3.4.3).

In order to be able to apply the comparison test to a series $\sum u_n$, it is clearly necessary to have some standard series, known to be either convergent or divergent, with which to compare $\sum u_n$. In this connection it is worth noting the following useful results:

(a) If $-1 < r < 1$, the geometric progression $\sum ar^n$ is convergent. (For $s_n = a/(1-r) - ar^n/(1-r) \to a/(1-r)$.)

(b) $1 + 1/2 + 1/3 + \ldots + 1/n + \ldots$ is divergent (see § 3.3).

(c) For $p > 1$, the series $1/1^p + 1/2^p + 1/3^p + \ldots + 1/n^p + \ldots$ is convergent (see § 3.7, the integral test).

(d) $1/1.2 + 1/2.3 + 1/3.4 + \ldots + 1/n(n+1) + \ldots$ is convergent. [Here s_n (by partial fractions) $= 1 - 1/(n+1) \to 1$.]

3.5. d'Alembert's Ratio Test

This test also may be expressed in either of two different forms. We assume at this stage that $u_n > 0$, and state the theorem as follows.

THEOREM A. *The series $\sum u_n$ is convergent if, for $n \geqslant N$, $u_{n+1}/u_n \leqslant k < 1$.*

Proof. We have $u_{N+2} \leqslant k u_{N+1} \leqslant k^2 u_N, \ldots, u_{N+p}, \leqslant k^p u_N$.
We can hence compare the series with the convergent geometric progression with common ratio k and $(N+p)$th term k^p, i.e. with

$$\sum v_n = k^{1-N} + k^{2-N} + \ldots + 1 + k + k^2 + \ldots .$$

We have $u_{N+p}/v_{N+p} \leqslant k^p u_N/k^p = u_N$. Hence, since u_N is fixed and $\sum v_n$, a geometric progression with common ratio $k < 1$, is convergent [see (a) in § 3.4 above], so also is $\sum u_n$ (by the comparison test).

THEOREM B. *For the series of positive terms $\sum u_n$, suppose that the limit of the ratio of consecutive terms, $\lim\limits_{n \to \infty} u_{n+1}/u_n = r$, exists.*
Then if $r < 1$, the series is convergent;
*   if $r > 1$, the series is divergent;*
*   if $r = 1$, the test is inconclusive.*

(It cannot be stressed too strongly that, in applying this test, the LIMIT of u_{n+1}/u_n, as n approaches infinity, is what is wanted. The ratio itself, if it contains n, is completely useless. For example, if we consider the divergent series $1 + 1/2 + \ldots + 1/n + \ldots$ the ratio $u_{n+1}/u_n = n/(n+1)$, which is certainly less than 1 for every n; but this fact does not imply the convergence of the series. In fact $n/(n+1) = 1 - 1/(n+1)$ which approaches 1 as n increases; and this value is the one that renders the test inconclusive.)

Proof. If $r < 1$, and if k is some number between r and 1 [say $(r+1)/2$, for example], then there exists an N such that, for $n > N$, we have $u_{n+1}/u_n < k$. For we have $|u_{n+1}/u_n - r| < \varepsilon$ for $n > N(\varepsilon)$, and we need merely choose ε less than $k-r$ $[= (1-r)/2$ in the case of k as defined above]. A diagram may help (Fig. 3.2).

All u_{n+1}/u_n for $n > N(\varepsilon)$ will then have values less than k, and the proof follows as in (a) above.

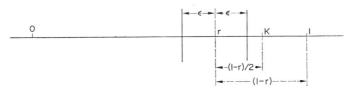

FIG. 3.2. d'Alembert's ratio test.

3.6. Upper and Lower Limits

DEFINITION. *Let* $\{u_n\}$ *be a sequence, not necessarily convergent, of real terms, for which a number L exists such that, for any* $\varepsilon > 0$, *and for* $n > N(\varepsilon)$, (i) *no* u_n *is greater than* $L + \varepsilon$, *but* (ii) *an infinite number of* u_n *exist such that* $u_n > L - \varepsilon$. *Then this number L is called the* upper limit *of* $\{u_n\}$ *as n tends to infinity. It is denoted by* $\overline{\lim} \, u_n$.

A similar definition exists for the lower limit ($\underline{\lim}$) of a sequence (see Exercise 3.6.1). It should be clear that if upper and lower limits of a sequence exist and coincide, then the sequence is convergent (Exercise 3.6.2). An unbounded sequence will evidently not have both upper and lower limits, and may have neither. For example, the sequence 1, 2, 1, 3, 1, 4, 1, 5, 1, ..., 1, n, 1, ..., has a lower limit ($= 1$) which also happens to be a lower bound. The sequence $\{u_n\}$, where $u_n = (1+1/n)-n$, has a least upper bound 1 but neither upper nor lower limit.

The sequence given by $u_n = (1+1/n)+(-1)^n$, which runs 1, $2\frac{1}{2}$, $\frac{1}{3}$, $2\frac{1}{4}$, $\frac{1}{5}$, $2\frac{1}{6}$, ..., has a least upper bound $2\frac{1}{2}$, an upper limit 2, and a greatest lower bound, which is also as it happens the lower limit, of zero.

It would obviously be possible for an infinite number of terms of a sequence to lie within an arbitrary ε of other values besides the upper and lower limits where they exist; for example, the sequence $\{u_n\}$, where $u_n = (1 + 1/n)\sin n\pi/4$, has an upper limit 1, a lower limit -1, and other limit points 0, $1/\sqrt{2}$, $-1/\sqrt{2}$.

In what follows, we do not pursue this matter further, being concerned only with the concept of upper limit.

3.7. Cauchy's Root Test

This test is of considerable importance in relation to the convergence of power series of complex terms.(These are series of form $\sum_{0}^{\infty} a_n(z - z_0)^n$, the a_n, z and z_0 being complex numbers. See Chapter 1.) We apply it here to the case of a series $\sum u_n$ of positive real terms.

Consider the sequence $\{u_1, u_2^{1/2}, \ldots, u_n^{1/n}, \ldots\}$, where in each case only the real positive root is to be taken. This sequence may have an upper limit as n tends to infinity. Let such a limit be L. Then:

THEOREM. *If $L < 1$, the series $\sum u_n$ is convergent;*
if $L > 1$, the series is divergent; and
if $L = 1$, the test is inconclusive.

The proof for convergence is similar to that of d'Alembert's ratio test in § 3.5, and is as follows.

Proof. If $L < 1$, let us select a k such that $L < k < 1$. Write $u_n^{1/n} = v_n$; then, for sufficiently large n, say $n > N$, we have $v_n < k$ (since v_n can be made to approach arbitrarily close to L, a smaller number than k). Thus for $n > N$ we have $u_n^{1/n} < k$ or $u_n < k^n$; and now, by the comparison test (Theorem A) above, comparing $\sum u_n$ with the geometric progression $1 + k + k^2 + \ldots + k^n + \ldots$, the result follows.

The proof of divergence when $L > 1$ is very easy, and is left as an exercise (3.7.1).

3.8. The Integral Test

(For this test some acquaintance with the ideas of function of a real variable, continuity, and integration is assumed. See also §§ 4.1 and 4.2.)

Let f be a continuous real function of a real variable x and let $f(x)$ be positive and monotonically decreasing for $x \geqslant 1$. Then we may frequently determine the convergence or divergence of $\sum_1^{\infty} f(n)$ by an examination of $\int_1^N f(x)\,dx$.

DEFINITION. *Let $f(x)$ be defined and continuous for all $x \geqslant a$. If a number L exists such that for any $\varepsilon > 0, \left| \int_a^N f(x)\,dx - L \right| < \varepsilon$ for $N > N_0(\varepsilon)$, then we say that $\int_0^{\infty} f(x)\,dx$ is* convergent, *and has the limit L. Otherwise the integral is said to be* divergent.

(Here N, N_0 represent real numbers, but not necessarily integers; and by $\int_a^N f(x)\,dx$ we mean $F(N) - F(a)$, where $d(F(x))/dx = f(x)$.)

THEOREM. *Let $f(x)$ be a positive monotonic decreasing function for $x \geqslant 1$. Then $\sum_1^{\infty} f(n)$ and $\int_1^{\infty} f(x)\,dx$ are either both convergent or both divergent.*

Proof. The theorem is best illustrated by means of a diagram (Fig. 3.3).

By consideration of the areas of rectangles of unit width "under" the integrable curve $y = f(x)$ ($1 \leqslant x \leqslant n$), we easily see that

$$f(1)+f(2)+\ldots+f(n-1) > \int_1^n f(x)\,dx > f(2)+f(3)+f(4)+\ldots+f(n),$$

which by an obvious notation we can rewrite as

$$S_n - f(n) > I_n > S_n - f(1).$$

From this we easily deduce that

$$f(n) < S_n - I_n < f(1),$$

and since $f(n) > 0$ it follows that $S_n - I_n$ is bounded above and below.

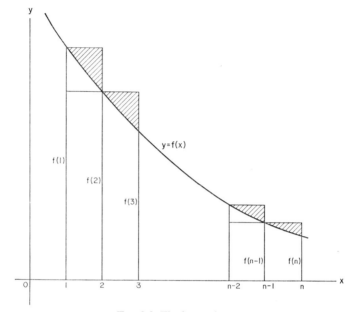

FIG. 3.3. The integral test.

Moreover, since $I_n > S_n - f(1)$ and both S_n and I_n increase with n, it follows that if as n tends to infinity I_n tends to a limit L, then $\{S_n\}$ is an increasing sequence with an upper bound $L + f(1)$, and thus $\{S_n\}$ converges (§ 2.17). On the other hand, since $S_n - I_n > 0$, it follows that if the integral diverges, i.e. increases without limit, then S_n also increases without limit, and the series diverges. Similarly, the convergence or divergence of the series implies that of the integral.

We consider the example [see § 3.4(c)] $S = \sum_{1}^{\infty} 1/n^p$ $(p > 0)$. Here $f(x) = 1/x^p$, which is positive and monotonically decreasing for $x \geqslant 1$, and the integral takes two forms according to whether or not $p = 1$. If $p = 1$ we have $\int_{1}^{N} dx/x = \log N$, which approaches infinity with N. Thus the integral diverges, and it follows that the series $S = \sum_{1}^{\infty} 1/n$ is divergent.

If p is not equal to 1, we have $\int_1^N dx/x^p = N^{1-p}/(1-p) - 1/(1-p)$,
which approaches infinity with N if p is less than 1; but if $p > 1$,
the value of the integral as N approaches infinity tends to $1/(p-1)$,
the first term tending to zero. Thus in this case, since the integral
converges, so does the series.

3.9. Series with Negative or Complex Terms

The above methods of testing for convergence apply only to series
of positive terms. When we come to deal with series containing negative or complex terms, special difficulties arise. In the case of negative
terms, it is not sufficient to consider the positive and negative terms
separately. For example, in the case of the series $S = 1 - \frac{1}{2} + \frac{1}{3} - \frac{1}{4} +$
$\ldots + (-1)^n 1/(n+1) + \ldots$ both the series

$$S_{(1)} = 1 + \frac{1}{3} + \frac{1}{5} + \ldots + 1/(2n+1) + \ldots \quad \text{and}$$
$$S_{(2)} = -\frac{1}{2} - \frac{1}{4} - \frac{1}{6} - \ldots - 1/2n - \ldots$$

are divergent, while the series S happens to be convergent, as will
shortly be shown. Another, at first sight curious, feature of a convergent infinite series containing both positive and negative terms
is that a rearrangement of the terms may, if it involves an infinite
number of terms, lead to a different sum. In the case considered above,
$1 - (\frac{1}{2} - \frac{1}{3}) - (\frac{1}{4} - \frac{1}{5}) - \ldots$, which on removal of brackets is seen to be
the series S, evidently has a sum less than 1 (if convergence is for the
moment assumed); while the rearrangement of the same terms to give
the series

$$T = 1 + (\frac{1}{3} + \frac{1}{5} - \frac{1}{2}) + (\frac{1}{7} + \frac{1}{9} - \frac{1}{4}) + \ldots,$$

in which each bracket is formed by two positive terms of S followed
by one negative one, is easily seen (assuming convergence) to have a
sum greater than 1, although all the terms of S, and no new ones, are
introduced.

This phenomenon is not as surprising as it at first appears to be,
since it must be remembered that we cannot in fact sum either series
by adding "all" the terms; we merely consider the limit of the nth

partial sum S_n, or T_n, as n increases, and for no value of n greater than 1 do the terms of the series correspond. Therefore there is no reason at all to expect $\lim_{n \to \infty} S_n$ and $\lim_{n \to \infty} T_n$ to be the same.

The following theorem is useful in relation to (real) series with alternating signs.

THEOREM. *If the terms of the sequence $\{u_n\}$ are real and positive, and monotonically decreasing to zero (i.e. $u_1 \geqslant u_2 \geqslant \ldots \geqslant u_n \geqslant \ldots \geqslant 0$, $\lim_{n \to \infty} u_n = 0$), then the series*

$$u_1 - u_2 + u_3 - u_4 + \ldots + (-1)^{n+1}u_n + \ldots \quad \text{is convergent.}$$

Proof. Consider first the sum to an even number of terms. We have

$$S_{2n} = (u_1 - u_2) + \ldots + (u_{2n-1} - u_{2n}), \quad \text{whence}$$
$$S_{2n+2} - S_{2n} = (u_{2n+1} - u_{2n+2}) \geqslant 0.$$

Thus S_{2n} is steadily increasing with n. Also

$$S_{2n} = u_1 - (u_2 - u_3) - \ldots - (u_{2n-2} - u_{2n-1}) - u_{2n} \leqslant u_1.$$

Hence S_{2n} is both monotonically increasing and bounded above, and thus converges to a limit ($\leqslant u$). (See § 2.17.)

Also $|S_{2n+1} - S_{2n}| = |u_{2n+1}| \to 0$ as $n \to \infty$, so that S_{2n+1} approaches the same limit as S_{2n}, and the given series is convergent.

The necessity for taking account of the sum to both an odd and an even number of terms, and for the condition $u_n \to 0$, is illustrated by the series

$$S = 3 - 2\tfrac{1}{2} + 2\tfrac{1}{4} - \ldots + (-1)^n \left(2 + \frac{1}{2^n}\right) + \ldots$$

in which the sum to $2n$ terms is greater than $\tfrac{1}{2}$, but less than 3, whence $\{S_{2n}\}$ is a convergent sequence since S_{2n} increases with n but remains less than 3. But $|S_{2n+1} - S_{2n}| > 2$ at each stage, so $\{S_n\}$ is not a Cauchy sequence (§ 2.15) and therefore does not converge.

3.10. Absolute Convergence

A more important test for convergence applicable to series with negative and also complex terms is based on the idea of absolute convergence.

DEFINITION. *If* $S = u_1 + u_2 + \ldots + u_n + \ldots$ *is a series of real or complex terms, and if* $T = |u_1| + |u_2| + \ldots + |u_n| + \ldots$, *then S is said to be* absolutely convergent *if T is convergent.*

It will be noted that T is a series of non-negative real terms; i.e. a series of a type to which we can apply the tests already considered. The importance of absolute convergence lies in the following key theorem.

THEOREM. *An absolutely convergent series of real or complex numbers is convergent.*

Proof. We have $|S_{n+p} - S_n| \leqslant |T_{n+p} - T_n|$, using the notation above and the result (§ 1.14 and Exercise 1.14.4), that the modulus of the sum $u_{n+1} + u_{n+2} + \ldots + u_{n+p}$ is less than the sum of the moduli, $|u_{n+1}| + \ldots + |u_{n+p}|$. But $\{T_n\}$ is convergent, and hence a Cauchy sequence. In other words, for a sufficiently large N we may make the right-hand side of the inequality less than any positive ε for $n > N$ and all $p \geqslant 1$. This establishes that $|S_{n+p} - S_n| < \varepsilon$ for $n > N$, whence $\{S_n\}$ is also a Cauchy sequence, and therefore convergent. This proves that the series S is convergent, which was what we wished to show.

The converse of this theorem is false, as can be seen at once by considering the series $S = 1 - \frac{1}{2} + \frac{1}{3} - \frac{1}{4} + \ldots$, which we know to be convergent, since the terms alternate in sign and decrease monotonically in magnitude to zero (§ 3.9). However, the series $1 + \frac{1}{2} + \frac{1}{3} + \ldots$ obtained by replacing each term by its modulus has been shown in § 3.3 to be divergent.

This theorem, that *an absolutely convergent series is convergent,* enables us to use the comparison test and the ratio test in investigating the convergence of series of complex numbers. When checking $\sum z_n$ for absolute convergence, the series we have to inspect is $\sum |z_n|$, which is obviously a series of non-negative terms; we are therefore in a situation in which the comparison test can be applied.

With regard to Theorem B of the ratio test, we note that if $\sum |z_n|$ is to be tested for convergence, we examine (if it exists) $\lim_{n \to \infty} |z_{n+1}/z_n| = r$, say. If $r < 1$ the series $\sum z_n$ is absolutely convergent, and hence convergent. If $r > 1$, for n greater than some N the terms are definitely increasing in magnitude, for we may choose a k such that $1 < k < r$, and then choose an ε so small (less than $r - k$ will suffice) that the ratio being within ε of r will ensure that it is greater than k. It then follows that $|z_{n+1}| > k|z_n| > |z_n|$. Thus (by § 3.3) the series $\sum z_n$ cannot converge, since the terms z_n are of continually increasing modulus and so cannot approach 0 as n approaches infinity.

If $r = 1$ the test is inconclusive. Other tests must be applied.

3.11. Other Tests

If the series is not absolutely convergent it may still, of course, be convergent. In the case of complex terms it is clearly necessary and sufficient for convergence that the series of real and imaginary parts each converge separately, i.e. $\sum (u_n + iv_n)$ is convergent if and only if $\sum u_n$ and $\sum v_n$ each converge. The proof is left as an exercise (3.11.1). (Compare § 2.14.)

Cauchy's root test, as indicated before, is of theoretical importance in relation to complex power series. Applied to the series of moduli $\sum |z_n|$ it is a test for the absolute convergence of $\sum z_n$. Consider the power series $\sum_{n=0}^{\infty} a_n(z - z_0)^n$, where z, z_0 and the a_r may be complex. The test states that

if $\overline{\lim} |a_n(z - z_0)^n|^{1/n} = k$, then $k < 1$ implies convergence,

$$k > 1 \text{ implies divergence,}$$

$$k = 1 \text{ provides no information.}$$

For convenience, we denote $\overline{\lim} |a_n|^{1/n}$ by $1/R$. Then, since $|(z - z_0)^n|^{1/n} = |z - z_0|$, the test reduces to the result that if $|z - z_0|/R < 1$ (i.e. if $|z - z_0| < R$) the series is absolutely convergent, while if $|z - z_0|/R > 1$ (i.e. if $|z - z_0| > R$) the series is divergent.

DEFINITION. *The circle* $|z-z_0| = R$, *which is such that the series*
$S = \sum\limits_{n=0}^{\infty} a_n(z-z_0)^n$ *is absolutely convergent at all points z within it,*
and divergent at all points outside, is called the circle of convergence
of the series S.

What happens at points on the circumference of the circle cannot be determined from this test, and is different for different series (see Exercises).

There are two special cases. If the series is convergent for all values of x, we say that the radius of convergence is infinite; and if the only point at which the series converges is z_0 itself (the series then being $a_0+0+0+0+\ldots$), we say that the radius of convergence is zero.

The determination of the radius R of the circle of convergence for a given series $\sum a_n(z-z_0)^n$ will not always require an application of the root test. In many cases d'Alembert's ratio test will serve instead. For $\lim\limits_{n\to\infty} |a_{n+1}(z-z_0)^{n+1}/a_n(z-z_0)_n| < 1$ or > 1 according as $|z-z_0|$ is less than or greater than $\lim\limits_{n\to\infty} |a_n/a_{n+1}|$ (assuming the limit to exist); whence by comparison with the above calculations we see that it follows that $\lim\limits_{n\to\infty} |a_n/a_{n+1}| = R$. As examples, consider the series:

(a) $S = 1+z+z^2+\ldots+z^n+\ldots$ Here, $R = 1$.

(b) $S = 1+z/2+(z/2)^2 \mid \ldots+(z/2)^n+\ldots$ $(R = 2)$.

(c) $S = 1+2!z+3!z^2+\ldots+(n+1)!z^n+\ldots$ $(R = 0)$.

(d) $S = 1+z+z^2/2!+\ldots z^n/n!+\ldots$ $(R \doteq \infty)$.

3.12. Multiplication of Series

Before leaving, for the present, the subject of infinite series, we state and prove an important result concerning the multiplication of one series by another.

As a preliminary, consider the product $S_n T_n$, where

$$S_n = u_1 + u_2 + \ldots + u_n, \quad T_n = v_1 + v_2 + \ldots + v_n.$$

We may form the product in several ways, among which we select the following possibilities:

$$(u_1v_1 + u_1v_2 + \ldots + u_1v_n) + (u_2v_1 + \ldots + u_2v_n) + \ldots$$
$$+ (u_nv_1 + u_nv_2 + \ldots + u_nv_n), \quad (1)$$

$$(v_1u_1 + v_1u_2 + \ldots + v_1u_n) + (v_2u_1 + \ldots + v_2u_n) + \ldots$$
$$+ (v_nu_1 + v_nu_2 + \ldots + v_nu_n), \quad (2)$$

or even

$$u_1v_1 + (u_1v_2 + u_2v_1) + (u_1v_3 + u_2v_2 + u_3v_1) + \ldots. \quad (3)$$

In (3) the technique is to select first all partial products such that the sum of the suffixes is the least possible, here 2; then those with suffix sum next lowest, 3; and so on. The last two groups will be $(u_{n-1}v_n + u_nv_{n-1})$ and the single term u_nv_n. Evidently with a finite number of terms involved it does not matter which method of computation is used, as all terms will be accounted for, and the sum will in each case give the product $S_n T_n$ of the sums of the separate series.

Suppose, however, that $S = \sum_1^\infty u_r$ and $T = \sum_1^\infty v_r$, where the series are convergent series of real, non-negative terms, and we wish to compute ST. In this case, none of the three processes above can be effectively carried out, since an infinite number of products is involved. In case (1) the multiplication of the v terms by u_1 can never be completed, and thus the multiplication by u_2 never started. In case (2) an infinite number of multiplications by v_1 has to be completed before a start can be made with v_2. However, in the case of the apparently more artificial procedure (3), initial terms in both u and v series are dealt with at an early stage, and terms with higher suffixes left until later. Therefore we can at least *define* a series, which we call the *product* of $\sum u_n$ and $\sum v_n$, by the method of (3). Schematically, such a procedure may be illustrated by the following diagram:

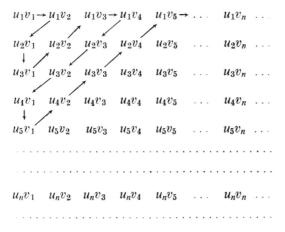

Each diagonal line indicates the members of one bracketed group of terms in (3), the order of addition being indicated by means of the arrows. Note that this differs slightly from (3) in that alternate bracketed groups have been added in reverse order; this is done merely in order to make it easier to recognize the pattern of the procedure.

We would like to know that this product series is convergent, and we hope that its limit will be *ST*.

If no terms with $n > N$ were to be included, the product $S_n T_n$ would be represented by the sum of all the terms in a square array, whereas the sum obtained by the "triangular" procedure above would, if only complete diagonal groups were included, take account of only about half of those terms. However, since the S and T series are both convergent, and the terms are non-negative, we note that

$$S_n T_n \leqslant C_{2n} \leqslant S_{2n} T_{2n},$$

where C_{2n} denotes the "triangular" sum (3) taken as far as the diagonal for which the sum of the suffixes is $2n$. This is because the partial products involved in $S_n T_n$ are all included in the triangle formed by the terms of C_{2n}, which are themselves all contained in the square formed by the terms of $S_{2n} T_{2n}$.

Moreover, since S_n approaches S and T_n approaches T as n increases, and S_{2n} and T_{2n} also approach these same limits, it follows that $S_n T_n$ and $S_{2n} T_{2n}$ both approach ST as n approaches infinity. Therefore C_{2n}, which lies between $S_n T_n$ and $S_{2n} T_{2n}$, must approach the limit ST also; which was what we wished to prove.

In the case of two absolutely convergent series of complex numbers the same rule of "multiplication" may be applied. The procedure is referred to as finding the "Cauchy product" of the series.

THEOREM. *If the series* $S = \sum_1^\infty u_n$ *and* $T = \sum_1^\infty v_n$ *are both absolutely convergent, then the series*

$$\sum_1^\infty (u_1 v_n + u_2 v_{n-1} + \ldots + u_{n-1} v_2 + u_n v_1)$$

converges to the sum ST.

Proof. Let $X_n = |u_1| + \ldots + |u_r| + \ldots + |u_n|,$
$$Y_n = |v_1| + \ldots + |v_r| + \ldots + |v_n|,$$

while $S_n, T_n,$ and C_n are defined as previously—with the difference that the series are now to consist of complex terms, not necessarily real non-negative terms.

We have　　　　$|C_{2n} - S_n T_n| \leqslant X_{2n} Y_{2n} - X_n Y_n,$

since the right-hand side of the inequality consists of the sum of the moduli of all terms on the left, and others besides. (It is helpful to inspect the diagram, considering the cases $n = 2$, $n = 3$.) But as n approaches infinity the right-hand side approaches zero, since $\{X_n\}$ and $\{Y_n\}$ are both by hypothesis convergent sequences. Hence $C_{2n} \to ST$.

We have not dealt with C_{2n+1}, but exactly the same argument applies. It is easily seen that all the terms of $S_n T_n$ are included in C_{2n+1}, and these terms themselves are all included in $S_{2n} T_{2n}$, and the proof follows as before. Hence $C_n \to ST$ as n approaches infinity, as was to be proved.

This theorem is of particular application to power series, of form $\sum_0^\infty a_r z^r$, $\sum_0^\infty b_r z^r$, for the method of multiplication "by triangles"

Flow diagram for testing convergence or divergence of the series Σz_n

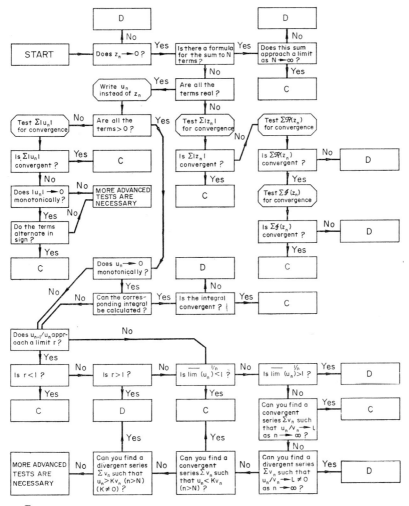

C: Σz_n is convergent
D: Σz_n is divergent

FIG. 3.4. Flow diagram *(see notes on next page)*.

Notes on using the flow diagram

(1) The flow diagram does not provide an exhaustive method for testing series for convergence. There is no mechanical procedure which will determine the convergence or divergence of every series.

(2) The procedure of following the flow diagram is more complicated than appears at first sight. For example, the instruction "Test $\sum|z_n|$ for convergence" means that the series $\sum|z_n|$ must itself be fed into the diagram at START.

(3) Whenever a series has some terms which are zero, consider instead the series obtained by removing all the zero terms. This makes no difference to the convergence or divergence properties and makes the series more amenable to treatment using the flow diagram (e.g. it becomes possible to consider u_{n+1}/u_n, which would not exist if $u_n = 0$).

(4) The convergence or otherwise of a series is unaffected by changing the signs of all the terms. Thus a series of negative terms can be changed into a series of positive terms. This procedure may occasionally be of use.

involves determining first the terms of lowest degree in z (possibly z^0), then those of next lowest degree, and so on, thus presenting the product series in its "natural" form as a third-power series in ascending powers of z.

We conclude this chapter with a (necessarily incomplete) "flow diagram" (Fig. 3.4) indicating how to test series systematically for convergence. Of course, in many cases most of the steps can be omitted, and the procedure that is recognized as most suitable can be applied at once.

While in this chapter we have considered only the very simplest tests for convergence, it should be pointed out that no system of testing exists which could be used to determine the convergence or otherwise of every conceivable series.

Exercises

3.1.1. Design series of real terms which, respectively:

(a) converge towards 2 from below;

(b) converge towards 2 from above (negative terms will be needed);

(c) converge towards 2 alternately from above and from below ($s_{2n}<2$, $s_{2n+1}>2$);

(d) do not converge but give partial sums always between 1 and 3;

(e) have partial sums which increase without limit;

(f) have an unbounded sequence of partial sums, yet are such that $\{s_{2n}\}$ is convergent.

Think of other possibilities and in each case devise more than one series to illustrate it.

3.1.2. Devise series of complex terms which, respectively:

(a) converge to a real number;

(b) converge to a complex number;

(c) have partial sums all of which have the same modulus, whilst amp s_{n+1} – amp s_n is constant;

(d) have partial sums represented by points lying along a fixed line (not one of the axes);

(e) diverge to infinity and are such that amp s_{n+1} – amp $s_n = \theta$, where θ is an angle which is not a multiple of $\pi/2$.

Consider other possibilities.

Note: in order to construct *series*, it is always useful to start with a *sequence* $\{s_n\}$, and then to let $u_n = s_n - s_{n-1}$ ($n > 1$), $u_1 = s_1$. Then clearly s_n is the partial sum $\sum_{r=1}^{n} u_r$.

3.2.1. (a) In the case of the non-convergent series $1-1+1-1+1-\dots$ calculate s_1, $(s_1+s_2)/2$, $(s_1+s_2+s_3)/3$, $\dots$, $(s_1+s_2+\dots+s_n)/n$, and observe whether these expressions tend to a limit as n increases.

"Sum" the series by the following device:

$$S = 1-(1-1+1-1+\dots) = 1-S, \text{etc.}$$

Compare your result with that obtained above, and that obtained in (§ 3.2). Note that the series, not being convergent, has no "sum" at all by convergence.

(b) Examine the series $1-2+3-\dots+(-1)^n(n+1)+\dots$.

Has this a sum by convergence? Justify your answer.

Attempt to obtain, by one or more of the methods suggested above, some number to represent the "sum" of the series. Consider whether you regard it as satisfactory to call this number the "sum" of the series.

3.3.1. Devise a convergent sequence $\{u_n\}$ such that $\sum_{1}^{\infty} u_r$ is convergent, and a convergent sequence $\{v_n\}$ such that $\sum_{1}^{\infty} v_n$ is divergent. State whether it is possible to find a divergent sequence $\{w_n\}$ such that $\sum_{1}^{\infty} w_n$ is convergent.

In the first of these cases, what can be said about the limit of the sequence $\{u_n\}$? State whether the existence of a limit taking this particular value is a sufficient guarantee of the convergence of $\sum u_n$. If not, give some counter-examples.

3.3.2. By the use of partial fractions, or otherwise, find the sum to n terms of the series $S = 1/1.3 + 1/3.5 + \dots + 1/(2n-1)(2n+1) + \dots$. Hence find the sum to infinity of the series.

3.3.3. Show that the series $S = 1 - \frac{1}{2} + \frac{1}{4} - \ldots + (-1)^n/2^n + \ldots$ is convergent· (Hint: find S_n.)

3.4.1. Prove that if $\sum v_n$ is a divergent series of positive terms, and if, for $n > N$, we have $u_n \geqslant k v_n$, where k is a positive constant independent of n, then $\sum u_n$ is divergent.
Prove that the series

$$S = 2 + \frac{3}{2} + \frac{2}{3} + \frac{3}{4} + \frac{2}{5} + \ldots + 2/(2n-1) + 3/2n + \ldots$$

is divergent.

3.4.2. In the previous example, compare the series with $\sum v_n$, where $v_n = 1/n$, and show that $\lim\limits_{n \to \infty} u_n/v_n$ does not exist.

3.4.3. Prove that if there are two series of positive terms $\sum u_n$ and $\sum v_n$, such that $\sum v_n$ is convergent, and $\lim u_n/v_n = 0$, then $\sum u_n$ is convergent. Prove also that if the ratio u_n/v_n approaches infinity with n, and if $\sum v_n$ is divergent, then $\sum u_n$ is also divergent. Comment on the case in which the ratio approaches zero, while $\sum v_n$ is divergent. Give an example of convergent series $\sum u_n$ and $\sum v_n$ such that $\lim\limits_{n \to \infty} u_n/v_n$ is infinite.

3.4.4. Determine whether the following series are convergent or divergent:

(a) $1 + \frac{1}{3} + \frac{1}{5} + \ldots + 1/(2n-1) + \ldots$;

(b) $1/1^2 + 1/2^2 + \ldots + 1/n^2 + \ldots$;

(c) $1000/1^2 + 1000/3^2 + \ldots + 1000/(2n-1)^2 + \ldots$;

(d) $(\frac{1}{2})/1^2 + (\frac{2}{3})/3^2 + (\frac{3}{4})/5^2 + \ldots + [n/(n+1)]/(2n-1)^2 + \ldots$;

(e) $(\frac{1}{2})/2 + (\frac{2}{3})/2^2 + \ldots + [n/(n+1)]/2^n + \ldots$.

3.5.1. Explain the reason for introducing k in the proof of the limit form of d'Alembert's ratio rest, § 3.5(b).

3.5.2. Test the following series for convergence or divergence:

(a) $1/3 + 2/3^2 + 3/3^3 + \ldots + n/3^n + \ldots$;

(b) $1 + 5/1! + 5^2/2! + 5^3/3! + \ldots + 5^n/n! + \ldots$;

(c) $1^2(\frac{1}{2}) + 2^2(\frac{1}{2})^2 + \ldots + n^2(\frac{1}{2})^n + \ldots$;

(d) $1^2(\frac{3}{4}) + 2^2(\frac{3}{4})^2 + \ldots + n^2(\frac{3}{4})^n + \ldots$;

(e) $1^2 + 2^2 + 3^2 + \ldots + n^2 + \ldots$.

Compare the results for (c), (d), and (e), and comment.

3.5.3. Determine for what real values of x the series $1 + x^2/3 + x^4/5 + \ldots + x^{2n}/(2n+1) + \ldots$ is convergent.

3.6.1. Give what you think is a reasonable definition for the lower limit of a sequence.

3.6.2. Prove that if the upper and lower limits of a sequence exist and coincide, then the sequence is convergent.

3.6.3. Give the upper and lower limits of the sequence $\{u_n\}$, where

$$u_n = 5 + \sin n\pi/6 + (-1)^n + \cos \pi/n.$$

3.6.4. Prove that an upper limit of a sequence, if it exists, is unique (i.e. show that if both L_1 and L_2 satisfy the conditions of L in the definition (§ 3.6), then $L_1 = L_2$).

3.6.5. Explain in your own words the difference between the idea of limit of a sequence and that of upper limit of a sequence.

3.6.6. The usual definition for the lower limit of a sequence $\{u_n\}$ is that it is equal to minus the upper limit (if it exists) of the sequence $\{-u_n\}$. Check whether your definition in Exercise 3.6.1 satisfies this.

3.6.7. Can the lower limit of a sequence be greater than the upper limit? Can the upper limit be greater than the l.u.b.? Give three examples in which the upper limit is (a) equal to, (b) less than the l.u.b.

3.7.1. If $\sum u_n$ is a series of real positive terms, if $v_n = u_n^{1/n}$, and if the sequence $\{v_n\}$ has an upper limit $L > 1$, prove that $\sum u_n$ is divergent.

3.7.2. Use the root test in order to determine for which positive values of x the following series converges:

$$S = 3^0 + 2x + 3^2 x^2 + 2^3 x^3 + \ldots + 3^{2n} x^{2n} + 2^{2n+1} x^{2n+1} + \ldots.$$

3.7.3. Given that $\lim_{n \to \infty} n^{1/n} = 1$, determine for what positive values of x the series

$$S = \sum_{1}^{\infty} (n+1)\,(x/4)^n \text{ converges.}$$

3.8.1. Prove in detail that (using the notation of § 3.8):

(a) if $\{S_n\}$ converges, then $\{I_n\}$ converges;

(b) if $\{S_n\}$ diverges, then $\{I_n\}$ diverges.

3.8.2. Investigate the convergence or divergence of $\sum_{n=0}^{\infty} 8/(2n+1)^p$, for real p.

3.8.3. Show that the series $\sum_{2}^{\infty} 1/n \log n$ is divergent.

3.8.4. Use the integral test in order to show that (a) $\sum_{1}^{\infty} n/(n^2+4)$ is divergent; (b) $\sum_{1}^{\infty} 1/(n^2+4)$ is convergent.

3.9.1. Devise some series which converge, but not absolutely; also some series which converge absolutely but for which d'Alembert's ratio test fails.

3.9.2. Determine whether each of the following series is convergent or divergent.

(a) $4-2+1-\frac{1}{2}+\ldots+(-1)^n 2^{2-n}+\ldots$;

(b) $4-3+2\frac{1}{2}-\ldots+(-1)^n(2+(\frac{1}{2})^{n-1})+\ldots$;

(c) $1-\frac{1}{3}+\frac{1}{2}-\frac{1}{9}+\ldots+1/2^n-1/3^{n+1}+\ldots$;

(Hint: calculate S_{2n}.)

(d) $1+\frac{1}{2}-\frac{1}{4}+\frac{1}{8}+\frac{1}{16}-\frac{1}{32}+\ldots+1/2^{3n}+1/2^{3n+1}-1/2^{3n+2}+\ldots$;

(e) $1+\frac{1}{2}-\frac{1}{3}+\frac{1}{4}+\ldots+1/(3n-2)+1/(3n-1)-1/3n+\ldots$;

(f) $\sin A+\sin 2A+\ldots+\sin nA+\ldots$ $(0<A<\pi)$;

(g) $1-i/2+i^2/4-i^3/8+\ldots+(-1)^n i^n/2^n+\ldots$.

3.10.1. Test the following series for convergence:

(a) $1+i/2+\ldots+i^n/2^n+\ldots$;

(b) $1/1.2+1/2.3-1/3.4+\ldots+1/(3n-2)(3n-1)+1/(3n-1)3n-1/3n(3n+1)+\ldots$;

(c) $\sin A+(\sin 2A)/2+(\sin 4A)/4+\ldots+(\sin 2^nA)/2^n+\ldots$.

3.10.2. Determine for what real values of x the following power series are convergent (the series is in each case $\sum u_n$):

(a) $u_n = x^n/n$;

(b) $u_n = x^n/n!$;

(c) $u_n = x^n/n(n+1)$;

(d) $u_n = (-1)^n x^{2n+1}/(2n+1)$.

3.10.3. Determine the values of $|z|$ for which the following series $\sum u_n(z)$ remain absolutely convergent:

(a) $u_n(z) = z$;   (b) $u_n(z) = z^n \sin nA/(n+1)(n+2)$ $(0<A<\pi)$.

In (b), discuss the case $A = \pi$.

3.10.4. Show that these series are convergent for $|z|<1$, and divergent for $|z|>1$. Examine the case $|z|=1$.

(a) $\sum n^2 z^n$;   (b) $\sum z^n$.

3.11.1. Prove that $\sum(u_n+iv_n)$ is convergent if and only if $\sum u_n$ and $\sum v_n$ are each convergent.

3.11.2. Verify the examples given at the end of § 3.11.

3.11.3. Determine the radius of convergence of each of the following series:

(a) $\sum (z-1)^n/2^n$;

(b) $\sum z^n/n(n+1)\,(n+2)$;

(c) $\sum (n+1)z^n/(2n+1)$;

(d) $\sum n!z^n/n^n$.

(Hint: $\lim\limits_{n\to\infty} (1+1/n)^n = e$.)

3.11.4. Examine the convergence of the following series when $z = 1, i, -1, -i$ respectively:

(a) $\sum z^n/n$;   (b) $\sum z^n/n^2$;   (c) $\sum (1+1/n^2)z^n$.

3.11.5. Determine whether the following series of complex terms are convergent or divergent:

(a) $\sum (3+2i)^n$;   (b) $\sum (1-i/100)^n$;   (c) $\sum (5-3i)^n/8^n$;   (d) $\sum n(1-i)^n$.

3.11.6. Find the region in the complex plane such that for all points (x, y) within it, the series $\sum ne^{nx}(\cos ny + i \sin ny)$ is convergent.

3.12.1. Write down the first few non-zero terms of the Cauchy product of the series for $\sin x$ and for $\cos y$ (see § 5.4). Hence write down the first few non-zero terms of the series for $\sin A \cos B + \cos A \sin B$ (starting with terms of lowest degree and grouping together terms of the same degree). Compare your result with the series for $\sin (A+B)$.

3.12.2. Prove that the rearrangement of the terms of an infinite series of positive terms does not alter the sum (if it exists). (Hint: let the first n terms of the series before rearrangement be included in the first $N\,(\geqslant n)$ terms of the rearranged series. Then $s_n \leqslant S_N \leqslant s$, where s_n and s refer to the original series, and S_N to the re-arranged series. Now let n increase.) This result is known as *Dirichlet's theorem*.

3.12.3. Show by examples that Dirichlet's theorem does not apply when there are an infinite number of both positive and negative terms. Consider the case of a finite number of one or the other.

3.12.4. If exp (z) is by definition the sum of the series $1+z+z^2/2!+z^3/3!+ \ldots +z^n/n!+ \ldots$, show, by comparing terms of the nth degree, that exp $(x+y) =$ exp (x) exp (y). (Here x and y represent complex numbers.)

CHAPTER 4

FUNCTIONS OF A COMPLEX
VARIABLE

4.1. The Definition of a Function

Let D be a subset of the complex plane (possibly the whole plane). Suppose that we have some rule which, for every z in D, gives us a complex number w—which may or may not be in D. We call such a rule a *function* defined on D, and using f to denote the function we write $f(z) = w$. The number w is referred to as the *value* of f at z, or, alternatively, as the *image* of z under f. The set D is called the *domain* of f.

As z varies over the whole of D, the complex number $w = f(z)$ will vary over some set of complex numbers E. This set E, which consists of all w such that $w = f(z)$ for some z in D, is called the range of f.

Of course two different values of z may give rise to the same w: $f(z_1) = f(z_2) = w_1$, but $z_1 \neq z_2$. If it so happens that *all* the values of z in D give rise to the same w, we say that f is a *constant* function. This is clearly equivalent to the range of f consisting of exactly one point.

A function f can sometimes be described by means of a single formula which enables $f(z)$ to be computed when z is given. For example, if $z = x+iy$, then the formulae

$$f(z) = 3x+2iy,$$
$$f(z) = x^2+y^2-2ixy,$$
$$f(z) = |x+iy|,$$
$$f(z) = x-iy \ (= \bar{z}),$$

78

all describe functions having the whole complex plane as domain. The values of these functions at $z = 3+4i$, for example, are respectively $9+8i$, $25-24i$, 5, and $3-4i$.

It often happens that a formula fails to be meaningful for points z lying in a certain subset S of the plane; in this case the formula defines a function whose domain is the complement of S, i.e. the plane with S removed. For example, $f(z) = 1/z$ defines a function on the plane with the origin $z = 0$ removed; $f(z) = 1/(1-|z|^2)$ defines a function on the plane with the unit circle $|z| = 1$ removed.

Strictly speaking, it is not essential, when a function is defined by means of a formula, to take as the domain all points for which the formula is meaningful; we shall, however, normally do so for our purposes. Other more limited domains will be specified when required.

To save unnecessary repetition, we shall use some shortened notation when referring to functions described by formulae. We shall call the formula itself the function, and shall assume the domain of the function to be the set of all points for which the formula makes sense. Thus we shall say "the function $f(z) = 1/(1+z^2)$" or "the function $w = 1/(1+z^2)$", or even "the function $1/(1+z^2)$", when we mean "the function f defined by the formula $f(z) = 1/(1+z^2)$, and with domain the whole complex plane excluding the points $\pm i$".

Sometimes a combination of formulae is used in order to describe a function on a domain where one formula is insufficient. For example, the combinations

$$\left. \begin{array}{ll} f(z) = (z^2-1)/(z^2+1) & (z \neq \pm i) \\ \quad\ = 0 & (z = \pm i) \end{array} \right\}$$

and

$$\left. \begin{array}{ll} f(z) = z & (|z| \leqslant 1) \\ \quad\ = 1/z & (|z| > 1) \end{array} \right\}$$

each give a function with the whole plane as domain.

Although all the points z and $w = f(z)$ lie in the complex plane, it is often very helpful when considering functions to think of two copies of the complex plane, one to contain the domain D of the function and the other to contain the range E (the set of all points w corresponding to points z in D). We shall refer to these planes as the z-plane and the w-plane respectively.

If we write $w = f(z)$ in the form

$$f(x+iy) = u(x, y) + iv(x, y)$$

where $u, v, x,$ and y are all real, we note that when we evaluate w, given z, we are in fact evaluating two real functions $u(x, y)$ and $v(x, y)$ of the two real variables x and y. It is thus natural to ask to what extent the theory of complex functions can be developed in a similar way to the theory of real functions of real variables.

The first property we shall investigate is that of continuity.

4.2. Continuity

DEFINITION. *A function f is said to be* continuous *at the point z_0 if for every positive number ε there is a positive number δ such that if $|z-z_0| < \delta$, then $|f(z)-f(z_0)| < \varepsilon$.*

This may be illustrated graphically (Fig. 4.1).

We take two complex planes, the z-plane and the w-plane. We plot z_0 in the z-plane and $w_0 = f(z_0)$ in the w-plane. With centre w_0 we construct a circle of radius ε. We can now assert that, if the function is continuous at z_0, then we can construct a circle with centre z_0, and

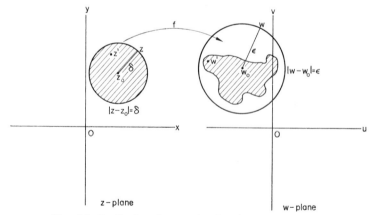

FIG. 4.1. Continuity of a complex function f at a point z_0.

radius δ, where δ is correctly chosen in relation to ε, such that if z' is any point inside the circle $|z-z_0| = \delta$, then the corresponding point $w' = f(z')$ lies within the circle $|w-w_0| = \varepsilon$. In general a smaller ε-circle will need a smaller δ-circle; continuity of f at z_0 means simply that a sufficiently small displacement of z from z_0 in any direction in the complex plane can at most produce a small change in the value of $w = f(z)$—the change in w may be made as small as we please by sufficiently restricting the displacement of z. The principle is exactly the same as that involved in the continuity of a real function of a single variable. A difference is that in the case of a function of one real variable the situation can easily be illustrated by a graph of the function, whereas in the complex case the z-plane and the w-plane need to be drawn separately, since the page is only two-dimensional, and a graph of f (i.e. $f(x+iy) = u(x, y)+iv(x, y)$) would require four dimensions to represent x, y, u, and v.

If f is continuous at all points of a given domain D it is said to be *continuous in D*.

We note that constant functions are evidently continuous everywhere. So also is the function f given by $f(z) = z$. Moreover (see Exercise 4.2.2) it is easily shown, as with real variable, that if f and g are continuous, at a point or in a domain, then so are $f+g$, $f-g$, fg [defined by $fg(z) = f(z)g(z)$], and f/g [provided that $g(z) \neq 0$]. It follows from repeated applications of these results that all polynomial functions (see § 5.1) are continuous, and so are functions of the form $P(z)/Q(z)$ where P and Q are polynomials, at points where Q does not vanish.

As a simple example of a continuous function consider the function f given by $f(z) = 3x-2iy$, where $z = x+iy$. If x and y are given (real) increments h and k respectively, the corresponding increment in w is easily seen to be $3h-2ik$. Thus the modulus of the increment in w is given by

$$|\Delta w| = +\sqrt{(9h^2+4k^2)}.$$

Let $\varepsilon > 0$ be arbitrary. If we now take $|h|$ as any number less than $\varepsilon/3 \sqrt{2}$ and $|k|$ as any number less than $\varepsilon/2 \sqrt{2}$, a simple calculation shows that this would make $|\Delta w|$ less than ε. But both $|h|$ and $|k|$ are less than or equal to $|\Delta z|$, the modulus of the increment in z, which is equal to $+\sqrt{(h^2+k^2)}$; so that if we take $|\Delta z| < \varepsilon/3\sqrt{2}$ we ensure that $|\Delta w| < \varepsilon$.

In the above example, chosen for simplicity, the Δw does not depend on z_0—merely on h and k. This is not generally the case (see Exercise 4.2.3). If, for a given ε, there is a value of δ which is applicable at all points of the domain D, the continuity is said to be *uniform*.

We note in passing that it is by no means necessary for *every* point inside $|w-w_0| = \varepsilon$ to be the image of a point inside $|z-z_0| = \delta$. [Consider the constant function $f(z) = w_0$, for example]. All we state is that *if* z satisfies $|z-z_0| < \delta$, then $|w-w_0| < \varepsilon$. In Chapter 7 we shall consider further the relationships between sets in the z-plane and their images in the w-plane under some well-behaved functions.

4.3. Differentiability

In real analysis, the process of differentiation is frequently illustrated by considering the gradient of the tangent to a curve $y = f(x)$ at a point (x, y). Any such intuitive interpretation of what might be meant by differentiation of a complex function is out of the question here—as we have already seen, a graph of the function $w = f(z)$ in the usual cartesian form is impossible, since four real variables, $x, y, u,$ and v are involved. Therefore we shall have to examine from first principles the definition of df/dx for a function of a real variable x, and decide whether or not a corresponding definition can be applied in the case of complex variables. We need first another definition.

DEFINITION. *Let f be a function of z defined in a neighbourhood of z_0 (although not necessarily at z_0 itself). Suppose that there exists a complex number L such that, given any $\varepsilon > 0$, there exists some $\delta > 0$ such that $|f(z)-L| < \varepsilon$ for all z with $|z-z_0| < \delta$. Then $f(z)$ is said to* approach L as z approaches z_0 and L is defined to be the limit *of $f(z)$ as z approaches z_0. We write $f(z) \to L$ as $z \to z_0$.*

(This definition obviously corresponds to the analogous one in real variable theory.)

Such a limit may not exist. For example as z approaches zero the function $1/z$ does not approach any number L. In cases where the limit exists, we may write

$$\lim_{z \to z_0} f(z) = L, \quad \text{or} \quad \lim_{h \to 0} f(z_0 + h) = L.$$

Note that if $f(z_0)$ exists, then to say that f is continuous at z_0 is exactly the same as saying that $f(z) \to f(z_0)$ as $z \to z_0$.

We now recall the definition of differentiability for a real function of a real variable.

DEFINITION. *If* $\lim_{h \to 0} [f(x_0+h)-f(x_0)]/h$ *exists, then we say that* f *is* differentiable *at* x_0. *The limit is called the* value *of the* derivative at x_0.

We may provisionally adopt a precisely similar definition for the differentiability of a complex function, as follows.

DEFINITION. *If* $\lim_{z \to 0} [f(z_0+z)-f(z_0)]/z$ *exists, then the complex function* z *is defined to be* differentiable *at* z_0, *and the limit is called the value of the derivative at* z_0, *and written* $df(z_0)/dz$, $(df/dz)_{z_0}$, *or* $f'(z_0)$.

Evidently this definition is applicable (trivially) to constant functions, since the limit as defined will be 0 in every case. Let us experiment with the function already considered, defined by $f(z) = 3x - 2iy$, which we have shown to be continuous. Fixing $z = x+iy$, and letting $Z = X+iY$ be a neighbouring (variable) point, we have, according to our provisional definition,

$$df/dz = \lim_{Z \to z} [(3X - 2iY) - (3x - 2iy)]/[(X + iY) - (x + iy)].$$

Let $X - x = h$, $Y - y = k$, then we require to consider the value of the above limit as h, k approach zero independently.

The expression easily reduces to $\lim_{h,\,k\,\to\,0} (3h-2ik)/(h+ik)$; but we now meet a difficulty, since no such limit exists. If, for example, we let $k = mh$, where m is some arbitrarily selected real constant, the expression reduces to $(3-2im)/(1+im)$, or, in $a+ib$ form, $(3-2m^2)/(1+m^2)$ $+i(-5m)/(1+m^2)$. Clearly the selection of different values for m will change the value of this expression. ($m = 0$, $df/dz = 3$; $m = 1$, $df/dz = \frac{1}{2}-2\frac{1}{2}i$). Since h and k are completely independent, it may seem at first that, apart from the trivial case of constant functions, it could not be expected that any complex function could have a derivative in the sense of the definition.

Consider, however, the function $w = A(x+iy)$, where A is a constant. According to our definition, if $w = f(z)$, then

$$df/dz = \lim_{h,\,k\,\to\,0} A(h+ik)/(h+ik) = A,$$

the entire factor $h+ik$ cancelling and a unique limit existing whatever the relationship between h and k. Thus certainly some functions exist which are differentiable in the sense of our definition, and we proceed to investigate what conditions are necessary and sufficient for differentiability.

4.4. The Cauchy–Riemann Equations

Let $w = f(z) = u(x, y)+iv(x, y)$,   where $z = x+iy$.

Then by definition, assuming the existence of df/dz, we have

$$df/dz = \lim_{h+ik\to 0} [f(z+h+ik)-f(z)]/(h+ik).$$

Consider first a displacement from z in the direction of the real axis. In this case $k = 0$ and we have to investigate

$$\lim_{h\to 0} \{[u(x+h, y)+iv(x+h, y)] - [u(x, y)+iv(x, y)]\}/h$$
$$= \lim_{h\to 0} [u(x+h, y)-u(x, y)]/h+i[v(x+h, y)-v(x, y)]/h$$
$$= \partial u/\partial x+i\,\partial v/\partial x.$$

Since we are assuming that df/dz exists, we shall require the same result if the displacement from z is in the direction of the imaginary axis. In this case, $h = 0$, and the calculation becomes

$$\lim_{k \to 0} [u(x, y+k) - u(x, y)]/ik + i[v(x, y+k) - v(x, y)]/ik$$

$$= (1/i) \, \partial u/\partial y + \partial v/\partial y.$$

If these results are to be the same, then by comparing real and imaginary parts of each expression we obtain

$$\partial u/\partial x = \partial v/\partial y, \quad \partial u/\partial y = -\partial v/\partial x,$$

these expressions being evaluated at (x, y).

These two equations are known as the *Cauchy–Riemann equations*. We have shown that they are a necessary condition for differentiability at $z = (x, y)$; but even if they hold we have no guarantee that different ways of approaching the point z, i.e. different relationships between h and k as they each approach zero, will not produce different results. We therefore examine whether or not the Cauchy–Riemann equations are sufficient conditions for differentiability.

4.5. The Cauchy–Riemann Equations. Sufficiency

Let us assume that $\partial u/\partial x$, $\partial u/\partial y$, $\partial v/\partial x$, and $\partial v/\partial y$ are continuous at all points z in the domain of f. Let $h + ik$ be a small change in $z \, (= x + iy)$. Let δu be the corresponding increase in u; i.e. we define

$$\delta u = u(x+h, y+k) - u(x, y).$$

Now

$$u(x+h, y+k) - u(x, y)$$

$$= u(x+h, y+k) - u(x, y+k) + u(x, y+k) - u(x, y)$$

which we may write in the form

$$(\partial u/\partial x)_{x, y+k} \cdot h + \alpha h + (\partial u/\partial y)_{x, y} \cdot k + \eta k,$$

or, by the continuity of $(\partial u/\partial x)$,

$$(\partial u/\partial x)_{x, y} + \beta)h + \alpha h + (\partial u/\partial y)_{x, y} \cdot k + \eta k,$$

where α, β, and η all approach zero with h and k [since, for example,

$$(\partial u/\partial y)_{x,y} = \lim_{k\to 0} [u(x, y+k)-u(x, y)]/k,$$

from which it follows that

$$u(x, y+k)-u(x, y) = (\partial u/\partial y)_{x,y}\cdot k+\eta k,$$

where η approaches zero as k approaches zero].

We can thus write

$$\delta u = [(\partial u/\partial x)_{(x,y)}+\varepsilon]h+[(\partial u/\partial y)_{(x,y)}+\eta]k, \qquad (1)$$

where ε and η both approach zero with h and k.

Similarly, we can write

$$\delta v = [(\partial v/\partial x)_{(x,y)}+\varepsilon']h+[(\partial v/\partial y)_{(x,y)}+\eta']k. \qquad (2)$$

Now in order to see whether or not df/dz exists we have to examine

$$\lim_{(h+ik)\to 0} \{f[(x+iy)+(h+ik)]-f(x+iy)\}/(h+ik)$$

$$= \lim_{(h+ik)\to 0} \{[u(x+h, y+k)+iv(x+h, y+k)]-[u(x,y)+iv(x,y)]\}/(h+ik)$$

$$= \lim_{h\to 0,\, k\to 0} (\delta u+i\delta v)/(h+ik).$$

Using eqns. (1) and (2) and the Cauchy–Riemann equations, the expression whose limit we wish to find may be written

$$[1/(h+ik)]\{(\partial u/\partial x)h-(\partial v/\partial x)k+i(\partial v/\partial x)h+(\partial u/\partial x)k\}+R, \qquad (3)$$

where $R = (\varepsilon h+\eta k+i\varepsilon' h+i\eta' k)/(h+ik)$.

We note that $|R| \leqslant |\varepsilon|+|\eta|+|\varepsilon'|+|\eta'|$, since $|h|$ and $|k|$ are each less than or equal to $|h+ik|$.

We require the limit as $h+ik$ approaches zero of the expression

$$[1/(h+ik)]\{h(\partial u/\partial x+i\partial v/\partial x)+ik(\partial u/\partial x+i\partial v/\partial x)\}+R$$

[a rearranged form of eqn. (3)] which is easily seen to be $\partial u/\partial x +i\partial v/\partial x$, since the limit of the other term R is evidently 0.

We observe that the limit is independent of the way in which $h+ik$ approaches zero. (Contrast this situation with the example

$f(z) = 3x - 2iy$ in § 4.3.) Therefore we have shown that, provided the partial derivatives are continuous, the Cauchy–Riemann equations are sufficient as well as necessary conditions for the differentiability of f.

The fact that continuity (which was used in showing that R approached zero) is necessary is shown by a counter-example (Exercise 4.5.1).

The question still arises whether any functions less trivial than the ones already considered $[f(z) = A, f(z) = A(x + iy)]$ satisfy the Cauchy–Riemann equations. We will construct such a function by assuming that $v = xy$, say, and trying to determine what functions u are such that u, v satisfy the Cauchy–Riemann equations. Then by the above u can be combined with v to give a differentiable function $f(z) = u + iv$.

If $v = xy$, we have $\partial v/\partial x = y$, $\partial v/\partial y = x$; whence, from the Cauchy–Riemann equations $\partial u/\partial y = -y$, $\partial u/\partial x = x$.

The first equation gives on integration

$$u = -y^2/2 + \phi(x),$$

where ϕ is some function of x alone. On substituting this expression for u in the second equation, we obtain

$$\phi'(x) = x$$

and thus

$$\phi(x) = x^2/2 + c,$$

where c is an arbitrary (real) constant.

Hence a set of differentiable functions $u + iv$, differing from each other only by a constant and with the given imaginary part $v = xy$, does exist. They are given by $f(z) = (\frac{1}{2}x^2 - \frac{1}{2}y^2) + ixy + c$, where c is a real constant. [Evidently if c were complex the expression would still be differentiable]. Thus non-trivial differentiable functions do exist.

A function which is differentiable in a neighbourhood of a point z_0 is said to be *regular*, or *analytic*, at z_0.

From this definition it follows that a function which is regular at z_0 is also regular at all points *in some neighbourhood of z_0*. Hence if we are given a function f which is regular at all points of some set D, we can find an *open* set D' containing D such that f is regular at all points of D'. In this case we say that *f is regular in (the domain) D'*.

Note that since df/dz is independent of the way in which h and k approach zero, it is frequently convenient to calculate it by taking k to be zero. This differentiation "in the x-direction" is precisely partial differentiation with respect to x.

In other words, if f is differentiable, then $df/dz = \partial f/\partial x$. For example, consider the function of the previous paragraph, $f(z) = \frac{1}{2}x^2 - \frac{1}{2}y^2 + ixy + c$. Differentiating with respect to x only, keeping y constant, gives $df/dz = \partial f/\partial x = x + iy$. We could equally, although it is usually less convenient, decide to differentiate in the y direction, keeping x constant. Since the increment in z is not now k, but ik, we shall now find that $df/dz = \partial f/i\partial y$; in fact $\partial f/\partial y = -y + ix = i(x + iy)$, whence $(1/i)\,\partial f/\partial y = x + iy$ as before. The point will become clearer on attempting to differentiate the function from first principles, taking first k and then h as zero.

4.6. Analytic Functions

We now obtain a result that will enable us to construct analytic functions with great ease, and also to decide at a glance whether certain types of function are analytic and in what domain.

Let $z = x + iy$ and $\bar{z} = x - iy$. Treat z and $\bar{z}$ as two independent variables, noting that $x = \frac{1}{2}(z + \bar{z})$, $y = 1/2i(z - \bar{z})$. We may now "differentiate" $w = f(z) = u + iv$ partially with respect to $\bar{z}$. (Although the numbers involved are complex, all the formal processes involved can be carried out just as with real numbers.) By the application of the "chain rule" for partial differentiation we obtain

$\partial f/\partial \bar{z} = \partial u/\partial \bar{z} + i\partial v/\partial \bar{z}$

$= \partial u/\partial x \cdot \partial x/\partial \bar{z} + \partial u/\partial y \cdot \partial y/\partial \bar{z} + i(\partial v/\partial x \cdot \partial x/\partial \bar{z} + \partial v/\partial y \cdot \partial y/\partial \bar{z})$

$= \partial u/\partial x \cdot (\frac{1}{2}) + \partial u/\partial y \cdot (-1/2i) + i[\partial v/\partial x \cdot (\frac{1}{2}) + \partial v/\partial y \cdot (-1/2i)]$

$= \frac{1}{2}(\partial u/\partial x - \partial v/\partial y) + (i/2)(\partial v/\partial x + \partial u/\partial y)$

$= 0$ if the Cauchy–Riemann equations apply.

This means that w is not explicitly a function of $\bar{z}$, but can be expressed in terms of z only. Thus if a regular function of z is expressible as an analytical formula, x and y occur in the formula only in the combination $x+iy$.

We note, for example, that the regular function which we have already encountered, $w = \frac{1}{2}x^2 - \frac{1}{2}y^2 + ixy + c$, which has the derivative $x+iy$, could simply have been written $w = \frac{1}{2}z^2 + c$. The derivative is, of course, z.

It is left as an exercise (Exercise 4.6.2) to prove that if f and g are differentiable functions of z, then so are $f+g$, fg, cf, where c is a complex constant, and $1/g$, where $g \neq 0$. From these results it is immediately deducible that polynomials are differentiable everywhere, and thus everywhere regular; and that rational functions $[f(z)/g(z)$ where f, g are polynomials] are regular at all points where the denominator does not vanish.

Moreover, the derivative of z^n may easily be verified to be nz^{n-1}. (Check this as an exercise. The details are exactly the same as in establishing that if, for real variables, $y = x^n$, then $dy/dx = nx^{n-1}$.) Therefore the formulae for the derivatives of polynomials and rational functions of a complex variable correspond to those for a real variable.

Finally, we observe that if f and g are both regular functions, then the function h defined by $h(z) = f\{g(z)\}$ is regular (at points where it is defined), and as in the case of real variable, we have

$$h'(z) = f'\{g(z)\} \cdot g'(z).$$

4.7. Laplace's Equation

If u and v are twice differentiable with respect to x and y, and all the partial derivatives are continuous, it follows at once from the Cauchy–Riemann equations that

$$\partial^2 u/\partial x^2 = \partial^2 v/\partial x \, \partial y$$

and

$$\partial^2 u/\partial y^2 = -\partial^2 v/\partial y \, \partial x$$

and hence that

$$\partial^2 u/\partial x^2 + \partial^2 u/\partial y^2 = 0.$$

Thus u satisfies what is called *Laplace's equation in two dimensions*. It is equally easy to show that v also satisfies this equation.

4.8. Orthogonal Families of Curves

Consider the curve $u(x, y) = c$, where u is the real part of a regular function f, and c a real constant. Differentiating with respect to x gives $\partial u/\partial x + \partial u/\partial y \cdot dy/dx = 0$. Thus if a point (x_0, y_0) lies on the curve, the gradient at that point is given by

$$(dy/dx)_{(x_0, y_0)} = -(\partial u/\partial x)_{(x_0, y_0)}/(\partial u/\partial y)_{(x_0, y_0)}.$$

There will (in general) be a curve of form $v(x, y) = k$, where v is the imaginary part of the function f, through the point (x_0, y_0), as is immediately seen by letting $k = v(x_0, y_0)$. The gradient of the tangent to this curve is given by

$$(dy/dx)_{(x_0, y_0)} = -(\partial v/\partial x)_{(x_0, y_0)}/(\partial v/\partial y)_{(x_0, y_0)}.$$

Using the abbreviated notation u_x for $\partial u/\partial x$, etc., and considering the values at (x_0, y_0) in each case we now have

$$(dy/dx)_u (dy/dx)_v = u_x v_x/u_y v_y,$$

where the u, v suffixes on the left indicate the curve concerned, and in virtue of the Cauchy–Riemann equations the right-hand side reduces to -1, showing that where a "u-curve" and a "v-curve" intersect, they do so at right angles (Fig. 4.2).

We may, in general, find precisely one u-curve and one v-curve through any point in the (x, y) plane. This means that by varying c and k we obtain two orthogonal families of curves, each of which is derived from a solution of Laplace's equation. Such orthogonal families play an important part in certain branches of physics, where they are met in the forms of lines of force and lines of equipotential, for example.

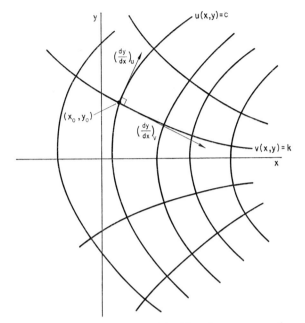

FIG. 4.2. Orthogonal families of curves.

It is evidently in some cases useful to be able to determine one family when the other is given. We dealt earlier (§ 4.5) with a simple case. Let us examine another example. Let $u = x^2$. We then have

$$\partial v/\partial y = \partial u/\partial x = 2x; \quad \partial v/\partial x - -\partial u/\partial y = 0.$$

These equations lead us in turn to

$$v = 2xy + \phi(x), \quad 2y + \phi'(x) = 0.$$

But the latter equation is impossible to satisfy except by taking y as constant. Thus we cannot in this case obtain a function $f(z) = u(x, y) + iv(x, y)$ of the form sought. What has gone wrong? The difficulty has arisen merely because the function $u = x^2$ is not a solution of Laplace's equation, which we have shown must be satisfied by any u or v which are the real or imaginary parts of a regular function.

When we have a u or v which satisfies Laplace's equation, it is still not always clear as to the best method of procedure in order to determine a regular function $w \, (= u + iv)$ even assuming it exists. One method which frequently suffices in easy cases is trial and error, using the knowledge that every x in u (or v) is part of an $x + iy$ in w. For example, if $u = x^2 - y^2$ (which does satisfy Laplace's equation) the obvious guess is $w = (x + iy)^2$, which does give the correct u, and leads to $v = 2xy$. [A more general solution is of course $w = (x + iy)^2 + ic$ (with c real).]

Another method is the one used already (cf.§ 4.5); if u is given, we find u_x and u_y, replace u_x by v_y, and u_y by $-v_x$, and re-integrate to find a suitable v. The procedure is to integrate v_x partially with respect to x, attaching to the integral, in place of an arbitrary constant, an arbitrary function of y, $\phi(y)$ say. The expression for v is now differentiated with respect to y, and the result compared with the known v_y. This will give an equation from which $\phi'(y)$, and hence $\phi(y)$ may be determined. Thus v can be completely determined.

A third method, which often proves useful when the procedure outlined above gives rise to difficult integrations, is given in Appendix C. It will be better understood when we have dealt with complex integration (Chapter 9).

Exercises

4.1.1. Calculate the values of the following functions of $z = x + iy$ at the points

(a) $z = 3 - 2i$;   (b) $z = 2 + i$;   (c) $z = 1 + 2i$.

(i) $f(z) = 2x + 3y$;   (ii) $f(z) = x - 2iy$;

(iii) $f(z) = x^2 - y^2 + 2ixy$;   (iv) $f(z) = (3x + 2iy)/(2x - 3iy)$,

giving your results in the form $p + iq$ (p, q, real) or p if real.

4.1.2. If $w = u + iv$ and $z = x + iy$, determine u and v in terms of x and y in the following cases:

(a) $w = z^3$;   (b) $w = (z - 1)/(z + 1)$;   (c) $w = 1/\bar{z}$;   (d) $w = 1 + z + z^2$.

4.2.1. For which values of z in the complex plane can the formula $w = (3z^3 - 2z^2 + 12z - 8)/(z^2 + 4)$ not be evaluated?

Show that the function f given by

$$f(z) = (3z^3 - 2z^2 + 12z - 8)/(z^2 + 4), \qquad z \neq \pm 2i \quad \left.\begin{array}{c} \\ \\ \\ \end{array}\right\}$$
$$= -2 + 6i, \qquad\qquad\qquad\qquad\qquad z = 2i$$
$$= -2 - 6i, \qquad\qquad\qquad\qquad\qquad z = -2i$$

is continuous everywhere in the complex plane.

4.2.2. Prove that if f and g are continuous functions of z in a given domain, then so are the functions $f + g$, fg, and $1/g$ (the last-named where $g(z) \neq 0$).

4.2.3. For the function $f(z) = (x + iy)^2$, determine the modulus of the variation in $|f(z)|$ produced by adding $0.1(1 + i)$ to z, when z is (a) 0, (b) $1 - i$, (c) $300 + 400i$.

Let $w = z^2$, and let the variation in w produced by a variation δz in z be δw so that $w + \delta w = (z + \delta z)^2$. Determine a single value, which need not be the smallest possible, of $|\delta z|$ which will ensure that $|\delta w| < 0.1$ whenever $|z| < 1000$.

4.2.4. If $w = x^2 + iy^2$, $z = x + iy$, find an s such that $|\delta z| < s$ implies that $|\delta w| < 1$ when $z = $ (a) 0; (b) $2 - i$; (c) $20 + 5i$.

4.3.1. If $w = x^2y(x + iy)/(x^4 + y^2)$, $z \neq 0$, and $w = 0$ when $z = 0$, prove that

(a) $\lim\limits_{z \to 0} w/z = \lim\limits_{z \to 0} (x^2y)/(x^4 + y^2)$;

(b) if $y = mx$, then $\lim\limits_{z \to 0} w/z = 0$ for all m;

(c) at the origin, $\delta w/\delta z$ approaches zero as z approaches zero along any straight line $y = mx$;

(d) If z approaches zero along the curve $y = x^2$, then $\lim\limits_{z \to 0} \delta w/\delta z = \frac{1}{2}$.

(e) $w = f(z)$ is not a differentiable function of z at the origin.

4.3.2. Investigate (from first principles) the differentiability (as a function of $z = x + iy$) of $w = x^3y(y - ix)/(x^6 + y^2)$ $(z \neq 0)$, $w(0) = 0$, at the origin.

4.3.3. Show that $w = xy$ is continuous at $z = 0$, and differentiable there, but not elsewhere, as a function of z.

4.4.1. Show that if $w = z^3$, then $dw/dz = 3z^2$ (a) from first principles, (b) by expressing in terms of x and y and using $dw/dz = \partial u/\partial x + i\, \partial v/\partial x$.

Experiment in this way with other expressions. Try differentiating in the y-direction.

4.4.2. Show that, for $z \neq 0$, $f(z) = (x - iy)/(x^2 + y^2)$ is a regular function of z.

4.4.3. Show that $e^x \cos y + ie^x \sin y$ satisfies the Cauchy–Reimann equations.

Test the following expressions to determine if, and in what region if so, they satisfy the Cauchy–Reimann equations.

(a) $\cos x \cosh y + i \sin x \sinh y$;

(b) $\cosh x \cos y + i \sinh x \sin y$.

Devise some more expressions with trigonometric or hyperbolic terms which satisfy the Cauchy–Riemann equations.

4.4.4. Show that $w = xe^x \cos y - ye^x \sin y + i(ye^x \cos y + xe^x \sin y)$ satisfies the Cauchy–Riemann equations everywhere in the complex plane.

4.5.1. Show that the function

$$w = (x^3 - y^3)/(x^2 + y^2) + i(x^3 + y^3)/(x^2 + y^2) \quad (z \neq 0), \quad w(0) = 0,$$

satisfies the Cauchy–Riemann equations at the origin. [Hint: take great care in differentiating partially with respect to x at the origin. Remember the definition of $u_x(0, 0)$.]

Show that the function is nevertheless not differentiable at the origin. [Hint: take the limit $w/(x + iy)$ as z approaches the origin along $y = mx$, or along the lines $y = 0$, $y = x$.] Explain the result. (See § 4.5.)

4.5.2. If $w = u + iv = x^2 - y^2 + 2ixy$, obtain values of $\varepsilon, \eta, \varepsilon', \eta'$ at the point $z = 1 + 2i$, as in eqns. (1) and (2) of § 4.5, given that $h = k = 0.2$.

4.6.1. In each of the cases (a) $w = z^n$ (n an integer); (b) $w = 1/(z^2 + 4)$; (c) $w = (z + 5)/(z - 5)$, determine any points in the complex plane at which w is not defined by the formula. Excluding these points, find in each case a formula for dw/dz.

4.6.2. Prove that if f and g are differentiable functions of z, then so are $f + g$, fg, and $1/g$ $[g(z) \neq 0]$.

4.6.3. Determine which of the following functions are regular over their domain of definition: $|z|$, amp z, e^{mz} ($\cos my + i \sin my$), $\mathcal{R}(z)$, $\mathcal{I}(z)$, $(3x - 3iy)^{-1}$, $(x^2 + y^2)/(3x - 3iy)$, $(1 - z^4)/(1 + z^4)$.

4.7.1. In each of the following cases show that u satisfies Laplace's equation: (a) $u = x^3 - 3xy^2$; (b) $u = e^{x^2 - y^2} \sin 2xy$; (c) $u = \sinh x \cos y$.

4.8.1. From the following equations, derive two orthogonal families of curves, identifying the nature of the curves involved. Sketch the families:

$$\text{(a) } w = z^2, \text{ (b) } w = 1/z, \text{ (c) } w = 1/(z + 1).$$

Derive other orthogonal families from some simple regular functions.

4.8.2. In each case, verify that the given u (or v) satisfies Laplace's equation; then determine a corresponding v (or u):

(a) $v = \cos x \sinh y$;

(b) $u = e^{x^3 - 3xy^2} \cos (3x^2 y - y^3)$;

(c) $v = -y/(x^2 + y^2 + 2x + 1)$;

(d) $u = y$;

(e) $u = \frac{1}{2} \log_e (x^2 + y^2)$ (in this example, identify the orthogonal families of curves);

(f) $u = 2xy$;

(g) $v = y(3x^2 - y^2)$.

CHAPTER 5

ELEMENTARY FUNCTIONS

IN THIS chapter we consider certain elementary functions of the complex variable $z = x + iy$. We have already shown that regular functions—the type with which we shall from now on be mainly concerned—contain x and y only in the combination $x + iy$ or z. In the functions considered in this chapter, $f(z)$ will be expressed explicitly in terms of z.

5.1. Polynomials

A polynomial in z, i.e. an expression of the form $a_0 + a_1 z + a_2 z^2 + \ldots + a_n z^n$, where the a_r may be complex numbers, may be differentiated term by term with respect to z [it is easily shown from first principles that $d/dz(z^r) = rz^{r-1}$ (see § 4.6)] and is regular over the whole complex plane. We note the following points. If $f(z) = a_0 + a_1 z + \ldots + a_n z^n$ is constant, then either $n = 0$, or $a_1 = a_2 = \ldots = a_n = 0$. If neither of these conditions hold, then $|f(z)|$ may be made as large as we please by sufficiently increasing $|z|$. It will be proved later that there are always values of z for which $f(z) = 0$, if we exclude the case $f(z) = \text{const}$. (See Chapter 10, Exercise 10.8.3.)

5.2. Rational Functions

These are functions of the form $f(z) = P(z)/Q(z)$, where P and Q are polynomials. The function is not defined at the zeros of Q, i.e. the values of z which make $Q(z)$ zero; as stated in the previous paragraph, these always exist unless $Q(z)$ is a constant. At all other points in the complex plane f is regular.

5.3. The Exponential Function

In real analysis, the function e^x, or as it is sometimes written for typographical or other reasons, exp x, is usually encountered at school level. In consequence certain difficulties of definition sometimes tend to be neglected, causing other difficulties later. For example, while a meaning can be attached to exp (p/q), namely the real positive qth root of e^p (assumed to exist), the problems raised in the interpretation of, for example, exp $(\sqrt{2})$ are often not discussed. The appeal to the index law $a^m a^n = a^{m+n}$, by which a meaning is evolved for $a^{p/q}$, is of no assistance here. One method of interpretation is based on the fact that the sequence exp (1), exp $(14/10)$, exp $(141/100)$, exp $(1414/1000)$, ..., constructed from the decimal approximations to $\sqrt{2}$, is convergent, so that exp $(\sqrt{2})$ may be defined as the limit of the sequence.

When we face the problem of the interpretation of exp z, where z is a complex number, this method cannot be used. Fortunately, however, another possibility exists.

It is well known that exp x can be expanded in the form of an infinite series:

$$\exp x = 1 + x + x^2/2! + \ldots + x^n/n! + \ldots,$$

and that this series could be taken as the definition of exp x. In the same way we *define* exp z to mean the series $1 + z + z^2/2! + \ldots + z^n/n! + \ldots$.

Naturally such a definition would only be applicable for regions in which the series converges; but the application of d'Alembert's ratio test easily verifies that it converges everywhere in the complex plane.

It should be noted that this definition, which is not based in any way on the index law, gives no reason to believe that the index laws apply to exp z; in particular, that exp $z \times$ exp $w = $ exp $(z+w)$. However, this result is actually true, and has already been met as Exercise 3.12.4. This allows us to carry out the usual manipulations with the index form.

Now we can also define a^z, where a is a real positive number; we can rewrite the expression in the form exp ($z \log a$) and expand; the series obtained will again converge everywhere in the complex plane.

5.4. Sine and Cosine

If θ is real, we may define $\sin \theta$ as the y-coordinate of the point of amplitude θ on the unit circle. Such a definition is obviously not applicable if we wish to attach a meaning to the expression $\sin z$, where z is complex.

In order to define $\sin z$, we again have recourse to a series expansion. Since we know that, when x is real,

$$\sin x = x - x^3/3! + x^5/5! - \ldots$$

we *define*

$$\sin z = z - z^3/3! + z^5/5! + \ldots + (-1)^n z^{2n+1}/(2n+1)! + \ldots.$$

Similarly, we *define*

$$\cos z = 1 - z^2/2! + z^4/4! + \ldots + (-1)^n z^{2n}/(2n)! + \ldots.$$

These series are again easily shown to be convergent over the entire complex plane, but, of course, such results as $\sin^2 z + \cos^2 z = 1$, $\sin 2z = 2 \sin z \cos z$ cannot be assumed to be true without further investigation. Such relations can be verified using the series above, but the work proves tedious. Fortunately, a surprising and useful relationship between the trigonometric series and the exponential series exists, which considerably simplifies our work with these functions.

5.5. The Link between the Exponential and Trigonometric Functions

Consider the expression

$$\exp (iz) = 1 + iz + \ldots + (iz)^n/n! + \ldots.$$

Since the series is absolutely convergent for all z, we may rearrange the terms and rewrite it in the form

$$\exp iz = (1 - z^2/2! + z^4/4! - \ldots) + i(z - z^3/3! + z^5/5! - \ldots)$$
$$= \cos z + i \sin z, \quad \text{for all } z.$$

Similarly,

$$\exp(-iz) = \cos z - i \sin z,$$

whence

$$\cos z = \tfrac{1}{2}(e^{iz} + e^{-iz}),$$
$$\sin z = (1/2i)(e^{iz} - e^{-iz}).$$

We now *define*

$$\tan z = \sin z / \cos z \quad (\cos z \neq 0)$$
$$= (e^{iz} - e^{-iz})/i(e^{iz} + e^{-iz})$$
$$= (e^{2iz} - 1)/i(e^{2iz} + 1),$$

and then $\cosec z = 1/\sin z$, $\sec z = 1/\cos z$, and $\cot z = 1/\tan z$, these functions being defined only where $\sin z$, $\cos z$, and $\tan z$ respectively are non-zero.

With these formulae it is easy to derive the elementary relations such as

$$\sin^2 z + \cos^2 z = 1,$$

$$\cos 2z = 2 \cos^2 z - 1$$

$$= 1 - 2 \sin^2 z,$$

$$\sin (w + z) = \sin w \cos z + \cos w \sin z$$

$$\sec^2 z = 1 + \tan^2 z,$$

and so on. (See Exercise 3.12.4.)

A result of particular interest and importance is obtained from letting $z = i\theta$, where θ is real. We then have

$$e^{i\theta} = \cos \theta + i \sin \theta.$$

Now, in our standard notation for complex numbers,

$$z = x + iy = r \cos \theta + ir \sin \theta$$
$$= r(\cos \theta + i \sin \theta)$$
$$= re^{i\theta}.$$

Hence we may write any complex number $z = x+iy$ in the "modulus–amplitude" form $re^{i\theta}$, where r, θ are real, and $r = |z|$, $\theta = $ amp z. This form frequently simplifies calculations; for example,

$$z_1 z_2 = r_1 e^{i\theta_1} r_2 e^{i\theta_2} = r_1 r_2 e^{i(\theta_1 + \theta_2)}$$

("multiply the moduli and add the amplitudes") in conformity with previous results.

Moreover, since the index law is applicable with complex indices (Exercise 3.12.4) we have in general

$$\exp(x+iy) = \exp x \exp iy$$
$$= \exp x (\cos y + i \sin y).$$

We shall see later (§ 5.11) that exp z, sin z, and cos z are regular over the whole complex plane.

5.6. de Moivre's Theorem

Another important result arises from expressing de Moivre's theorem (§ 1.15) in **exponential** form. We have

$$e^{2k\pi i} = \cos 2k\pi + i \sin 2k\pi$$
$$= 1 \quad (k = 0, \pm 1, \pm 2, \ldots).$$

Thus $e^{z+2k\pi i} = e^z$, from which we easily obtain

$$e^{z/n} = e^{x/n + i(y+2k\pi)/n}.$$

For, suppose $z_1^n = e^z$; then if $z_1 = x_1 + iy_1 = r_1 e^{i\theta_1}$ we have

$$r_1^n e^{in\theta_1} = e^z = e^x e^{i(y+2k\pi)}.$$

Hence $r_1 = e^{x/n}$ and $\theta_1 = (y+2k\pi)/n$, where $k = 0, \pm 1, \pm 2, \ldots$.

Many such results which have already been obtained in Chapter 1 are more compactly expressed by means of this notation, which we shall from now on adopt.

5.7. Hyperbolic Functions

Hyperbolic functions cosh z, sinh z, etc., may be readily defined as for real variables; we *define*

$$\cosh z = \tfrac{1}{2}[\exp z + \exp(-z)], \quad \sinh z = \tfrac{1}{2}[\exp z - \exp(-z)],$$
$$\tanh z = [\exp z - \exp(-z)]\,[\exp z + \exp(-z)]$$
$$= (\exp 2z - 1)/(\exp 2z + 1),$$
$$\operatorname{sech} z = 1/\cosh z \quad (\cosh z \neq 0), \text{ etc.}$$

We note that

$$\sin iz = i \sinh z, \quad \cos iz = \cosh z,$$
$$\sinh iz = i \sin z, \quad \cosh iz = \cos z,$$

from which relationships many hyperbolic formulae may be readily obtained from the corresponding trigonometrical formulae. It is also easy to obtain expressions for sinh z, cosh z in the form of power series by combining the series for exp z, exp $(-z)$.

5.8. The Logarithmic Function

If x and y are real and $e^x = y$, then we define x to be $\log_e y$, or, briefly, $\log y$, where x is unique if y is positive, and non-existent otherwise. The situation when a complex index is involved is more complicated. In fact, if $e^w = z$, there are an infinite number of possible values of w for any given z $(z \neq 0)$.

In order to see this, let $w = a + ib$ and $z = x + iy$; we have

$$e^a(\cos b + i \sin b) = x + iy,$$

whence, by equating real and imaginary parts,

$$e^a \cos b = x, \quad e^a \sin b = y.$$

Squaring and adding

$$e^{2a} = x^2 + y^2 = r^2, \quad \text{where } r = |z|,$$

giving

$$e^a = r, \quad a = \log r \text{ (uniquely)}, \quad \text{since } e^a > 0.$$

Also, $\cos b = x/r$, $\sin b = y/r$, and it follows that $b = $ amp z, where we may take *any* amplitude of z, since the addition of $2n\pi$ to the principal amplitude will not alter the sine or cosine of the angle.

Thus if $z = re^{i\theta}$ and $e^w = z$, then $w = \log r + i\theta + 2n\pi i$ ($n = 0, \pm 1, \pm 2, \ldots$). We will use the form Log z to denote the general logarithm of this form, and reserve the form log z to denote $\log r + i\theta$, where θ is the principal amplitude of z. We shall refer to the latter form as the *principal logarithm* of z. It gives w as a function of z in the sense that a unique value of w results from any given $z \neq 0$. As examples:

$\log 1 = 0$, $\quad$ Log $1 = 2n\pi i$,

$\log(-1-i) = \frac{1}{2}\log 2 - 3\pi i/4$, Log $(-1-i) = \frac{1}{2}\log 2 - 3\pi i/4 + 2n\pi i$,

$\log(3+4i) = \log 5 + i\tan^{-1}\frac{4}{3}$, Log $(3+4i) = \log 5 + i\tan^{-1}\frac{4}{3} + 2n\pi i$,

$\log(3-4i) = \log 5 + i\tan^{-1}(-\frac{4}{3}) = \log 5 - i\tan^{-1}(\frac{4}{3})$,

$\log(-3+4i) = \log 5 + i[\pi - \tan^{-1}(\frac{4}{3})]$,

$\log(-3-4i) = \log 5 + i(\tan^{-1}(\frac{4}{3}) - \pi)$.

It is important to note that Log is not, strictly speaking, a *function*. A function f by definition assigns one complex number $w = f(z)$ to each z in its domain, whereas Log (z) consists of an *infinite set* of complex numbers (whose imaginary parts differ from each other by integral multiples of 2π) for each non-zero z. The term "many-valued function"† is sometimes used in this context.

In dealing with these many-valued functions it is usually convenient to consider them as collections of genuine (i.e. "single-valued") functions, each defined on some suitably chosen domain. For example, we could consider Log (z) as the collection $\{\log z, \log z + 2\pi i, \log z - 2\pi i, \log z + 4\pi i, \log z - 4\pi i, \ldots\}$, where each member of the collection is a function defined on the whole plane except for the origin. In this case, however, none of the functions are continuous at points on the negative real axis [see Exercise 5.8.5(a)]. We can improve the situation by considering Log (z) instead as the collection

$$\{\log z + 2n\pi i, n = 0, \pm 1, \pm 2, \ldots\}$$

† Also "multiple-valued fuction".

each defined on the plane minus the negative real axis, together with the collection

$$\{\log(-z)+(2n+1)\pi i, \; n = 0, \; \pm1, \; \pm2, \; \ldots\}$$

each defined on the plane minus the positive real axis: in this case each of the functions is continuous and in fact regular in its domain of definition (see § 5.12). Such functions are called regular *branches* of the original many-valued function.

Let C be a closed path around the origin and let z be a point on C. Let $w(z)$ be any chosen value of Log (z): thus $w(z) = f(z)$ for some branch f of Log, where the domain of f contains z. As z goes around C, we may choose $w(z)$ to vary continuously with z, although this will entail using more than one branch of Log (owing to the exclusion of points on the negative real axis from one collection, and points on the positive real axis of the other collection, of branches of Log z). It is easy to verify that when z returns to its original position we have $w(z) = f_1(z)$, where f_1 is another branch of Log with $f_1(z) = f(z)+2\pi i$, the sign depending on the sense in which z encircled the origin [Exercise 5.8.5(b)]. This is the case no matter how small the path C around the origin may be. The origin is called a *branch-point* of the many-valued function Log.

As another example of a many-valued function, consider $\sqrt{z}$. For each $z \neq 0$ this has two values given by $r^{1/2}e^{(1/2)i\theta}$ and $r^{1/2}e^{(1/2)i\theta+i\pi}$, where $z = re^{i\theta}$ and θ is the principal value of amp z. On the plane minus the negative real axis we have two regular branches f_1, f_2 given by $f_1(z) = |z|^{1/2} e^{(1/2)i\theta}$, $f_2(z) = |z|^{1/2}e^{(1/2)i\theta+i\pi} = -|z|^{1/2}e^{(1/2)i\theta}$, and on the plane minus the positive real axis we have two regular branches given by

$$g_1(z) = |z|^{1/2}e^{(1/2)i\phi},$$
$$g_2(z) = |z|^{1/2}e^{(1/2)i\phi+i\pi} = -|z|^{1/2}e^{(1/2)i\phi},$$

where ϕ is the principal value of amp $(-z)$. It is easy to check that as z circles the origin once, a value $w(z)$ of $\sqrt{z}$ which varies continuously will not regain its original value when z does. On the other hand, if z describes a circle not surrounding the origin and returns to its original value, then so does $w(z)$, since there is no change of 2π in the amplitude of z. Therefore the origin is the unique branch-point for $\sqrt{z}$.

Similarly, for $a \neq 0$ the many-valued function $\sqrt{(z-a)}$ has precisely one branch-point, namely the point a.

5.9. More General Power Functions

Further consideration can now be given to the expression Z^z, where Z and z are both complex.

If a is real then $a^z = \exp(z \log a)$ by definition. Hence we shall *define*

$$w = Z^z = \exp(z \operatorname{Log} Z)$$
$$= \exp(x+iy)(\log R + i\Theta + 2n\pi i) \quad (n = 0, \pm 1, \pm 2, \ldots),$$

where $Z = R \exp i\Theta$.

This may be written in the form $\exp(A+iB)$, where

$$A = x \log R - y\Theta - 2n\pi y,$$
$$B = y \log R + x(\Theta + 2n\pi) \quad (n = 0, \pm 1, \pm 2, \ldots),$$

giving

$$w = \exp(x \log R - y\Theta - 2n\pi y) \times$$
$$\times [\cos(y \log R + x(\Theta + 2n\pi)) + i \sin(y \log R + x(\Theta + 2n\pi))].$$

Evidently we may expect this expression in general to take an infinite number of values for a given z. See Exercise 5.9.1 for further consideration of this matter. We can, however, define a principal value of Z^z simply by taking the principal logarithm of Z; then $n = 0$ and the formula for the principal value of Z^z (where $Z = Re^{i\theta}$ and $z = x+iy$) is given by

$$Z^z = \exp(x \log R - y\Theta)[\cos(y \log R + x\Theta) + i \sin(y \log R + x\Theta)].$$

5.10. The Expression of a Regular Function as a Series

Several of the functions considered above (but not all of them) were defined by means of infinite series. We shall find later (Chapter 11) that a function of z which is regular at a given point a in the z-plane may always be expressed as a series of non-negative powers of $(z-a)$

in the neighbourhood of that point. We shall frequently find it convenient to deal with regular functions expressed in that form. On the other hand, the fact that a function is defined by means of such a series does not yet prove that it is regular. We need to demonstrate differentiability. This involves the use of a very important theorem which we now prove.

5.11. Differentiability of Power Series

THEOREM. *If* $f(z) = \sum_{0}^{\infty} a_n z^n$ *for* $|z| < R$, *then* $f'(z)$ *exists for* $|z| < R$, *and* $f'(z) = \sum_{1}^{\infty} na_n z^{n-1}$.

Proof. We first prove some preliminary results.
(1) If $|w| < 1$, then

$$\sum_{0}^{\infty} w^n = 1/(1-w).$$

This may easily be proved either by developing the right-hand side by long division and considering the remainder after n terms, or by summation of the geometrical progression.
(2) If $|w| < 1$, then

$$\sum_{1}^{\infty} nw^{n-1} = 1/(1-w)^2.$$

If we denote the left-hand side by s, and consider $s - ws$, which reduces to

$$\sum_{0}^{\infty} w^n = 1/(1-w),$$

the result follows immediately. Alternatively, by applying Cauchy's method in order to "square" the series for $1/(1-w)$ the result is easily verified.

Of course, both these series are simple examples of the application of the binomial theorem with a complex variable; but we have not proved this theorem, and we are therefore not justified in using it in order to derive these results. We may briefly note, however, that the binomial theorem can be applied in the usual way to $(1+w)^n$, where n is real and $|w| < 1$. The convergence of the resulting series is immediately established by means of d'Alembert's ratio test.

In fact we shall only use the above two results in the case when w is real and positive.

(3) If

$$f(z) = \sum_0^\infty a_n z^n$$

is convergent, then some number K exists such that $|a_n z^n| < K$ for every n. This follows at once from the fact that the individual terms of the series form a sequence which converges (to zero) and is thus bounded (§ 2.13).

We are now in a position to prove the main theorem. The technique to be adopted is one of considerable importance at a later stage—we examine the difference between the result expected and that actually obtained, and show that it is in fact zero; thus establishing the theorem.

By definition,

$$df(z)/dz = \lim_{h \to 0} [f(z+h)-f(z)]/h.$$

In what follows, we shall assume that

$$|z| = r < \varrho < R,$$
$$|h| = m,$$

where $r+m < \varrho$. This will ensure that the points z, $z+h$, each lie within the circle of convergence $|z| = R$.

Thus

$$[f(z+h)-f(z)]/h = \left[\sum_{h=0}^\infty a_n(z+h)^n - \sum_{n=0}^\infty a_n z^n \right] \Big/ h$$

$$= \sum_{n=1}^\infty a_n[nz^{n-1}+n(n-1)z^{n-2}h/2!+\ldots+h^{n-1}].$$

If differentiation term by term is applicable, we shall expect, as we let h approach zero, that only the terms $\sum a_n n z^{n-1}$ will remain on the right-hand side. We thus examine the remaining terms which we will denote by E.

We have

$$E = \sum_{1}^{\infty} a_n[n(n-1)z^{n-2}h/2! + \ldots + h^{n-1}]$$

from which it follows that

$$|E| < \sum_{1}^{\infty} (K/\varrho^n)\,[n(n-1)r^{n-2}\,m/2! + \ldots + m^{n-1}]$$

(cf. paragraph (3) above and also Exercise 1.14.5).

Now the expression in square brackets may plainly be written in the form

$$(1/m)\,[(r+m)^n - r^n - nr^{n-1}m]$$

since, apart from a substitution of symbols, it has just been obtained from an expression of this kind. Hence

$$|E| < (K/m)\sum_{0}^{\infty} [(r+m)^n/\varrho^n - r^n/\varrho^n - nr^{n-1}m/\varrho^n].$$

The two geometrical progressions and the arithmetico-geometric progression may be summed to infinity as we noted at the beginning of this proof [see preliminary results (1) and (2) above]. This gives

$$|E| < (K/m)\left[1\Big/\left(1 - \frac{r+m}{\varrho}\right) - 1/(1-r/\varrho) - (m/\varrho)\,[1/(1-r/\varrho)^2]\right]$$

$$= (K/m)\,[\varrho/(\varrho-r-m) - \varrho/(\varrho-r) - m\varrho/(\varrho-r)^2].$$

Taking out a common factor ϱ and combining the first two terms in the main bracket gives

$$|E| < (K\varrho/m)\,[m/(\varrho-r-m)\,(\varrho-r) - m/(\varrho-r)^2]$$

which further reduces, on dividing out by m and bringing to a common denominator, to

$$|E| < K\varrho m/(\varrho-r-m)\,(\varrho-r)^2.$$

Evidently as m approaches zero, so does F, and the result is proved that if $f(z) = \sum_0^\infty a_n z^n$, then for $|z| < R$, $f'(z)$ exists and $= \sum_1^\infty a_n n z^{n-1}$.

Notes. (1) In this proof we have tacitly assumed the convergence of

$$\sum_u^\infty n a_n z^{n-1}.$$

This point is covered in the next section.

(2) There is no difficulty in extending the above theorem to the more general power series

$$F(z) = \sum_0^\infty a_n (z-z_0)^n, \quad |z-z_0| < R.$$

(3) By means of this theorem we easily show that $\exp z$, $\sin z$, and $\cos z$ are regular over the entire complex plane, and in particular that

$$\frac{d}{dz}(\exp z) = \exp z,$$

$$\frac{d}{dz}(\sin z) = \cos z,$$

$$\frac{d}{dz}(\cos z) = -\sin z.$$

5.12. Repeated Differentiation of an Infinite Series

After differentiation we have a second infinite series which can again be differentiated within its own radius of convergence; so it becomes of importance to know how the latter radius of convergence is related to that of the original series. The radius of convergence R, of $\sum_1^\infty a_n z^{n-1}$ is given by

$$1/R_1 = \varlimsup_{n \to \infty} |n a_n|^{1/n} \quad (\text{cf. } \S \ 3.10).$$

But $\lim_{n \to \infty} n^{1/n} = 1$, whence it easily follows that R_1 is equal to R, the radius of convergence of the original series. This immediately leads to the fact that a power series may be differentiated term by term as many times as desired within its radius of convergence.

On the circumference of this circle differentiation is not necessarily possible (see Exercise 5.11.4). Meanwhile we have established that the sum of a power series of complex terms is a regular function within the circle of convergence of the series. It follows easily that sin z, cos z, exp z, sinh z, and cosh z are regular everywhere. Log z cannot be expressed as a power series in z, and is not defined at zero; at other points, Log $z = \log z + 2n\pi i$ is easily seen to be regular in a suitable "cut plane" for each fixed n if we use the inverse relationship $z = e^w$ in order to obtain $dw/dz = 1/z$.

If $Z = f(z)$ is a regular function of z in some domain D, and if

$$w = Z^z = \exp (z \log Z),$$

then

$$dw/dz = \exp (z \log Z) \cdot \left(\log Z + \frac{z}{Z} \cdot \frac{dZ}{dz} \right).$$

Thus w is a regular function except where $Z = 0$ in D.

5.13. Inverse Functions

Exactly as with real variable we can define many-valued functions $\text{Sin}^{-1} z$, $\text{Tan}^{-1} z$, $\text{Cosh}^{-1} z$, etc., and then by restricting ourselves to the principal value obtain single-valued functions $\sin^{-1} z$, etc., regular except at easily identifiable points in the complex plane. Some of the standard results are included as exercises (5.13.1).

Exercises

5.1.1. If $w = z^3 + 4z$:

(a) find all points in the complex plane where $w = 0$;
(b) show from first principles that w is continuous at every point in the z-plane;
(c) calculate dw/dz from first principles;
(d) establish that w is everywhere regular.

5.1.2. Prove that if $f(z) = a_0 + a_1 z + \ldots + a_n z^n$, where the a_r may be complex, then for any $N > 0$ points exist in the z-plane such that $|f(z)| > N$. (Hint:

$$|f(z)| \geqslant a_n z^n| - |a_{n-1} z^{n-1}| - \ldots - |a_0|$$
$$= |z|^n(|a_n| - |a_{n-1}/z| - \ldots - |a_0/z^n|, \quad \text{etc.})$$

5.2.1. Determine all points in the z-plane at which the function $w = (z^2 - 4)/(z^2 + 4)$ is not defined. Show that the function is regular at all other points in the z-plane and evaluate dw/dz at the points $z = 1, 2, i, -i, 1+2i$.

5.3.1. Evaluate $\exp \sqrt{2}$ correct to two decimal places (a) by using the series expansion; (b) by means of logarithms.

5.3.2. Verify by multiplication of series that $(\exp z)^2 = \exp(2z)$.

5.3.3. Find the first few terms of the series expansion of 3^{1+i}, and reduce your expression to the form $a + ib$.

5.3.4. Prove that for no value of z can $\exp z = 0$. [Consider $\exp z \times \exp(-z)$.] Are there any other values that $\exp z$ cannot assume? (See § 5.8.)

5.3.5. Multiply $\exp z$ by $\exp(-z)$ using the series definition of the functions.

5.3.6. Using series definitions, verify that $\exp 3i \div \exp 2i = \exp i$. (Hint: cross-multiply.)

5.4.1. Prove the sine and cosine series to be convergent for all complex z.

5.4.2. Verify by direct multiplication of series as far as the term in z^6 that $\sin 2z = 2 \sin z \cos z$ (to at least this term).

5.4.3. Defining $\tan z$ as $\sin z/\cos z$, determine by direct division the first three non-zero terms of a power series for $\tan z$.

5.5.1. Prove the following formulae, in which z and w are complex numbers:

(a) $\sin^2 z + \cos^2 z = 1$.

(b) $\cos(z+w) = \cos z \cos w - \sin z \sin w$.

(c) $\tan 2z = 2 \tan z/(1 - \tan^2 z)$.

(d) $\cos z - \cos w = 2 \sin \frac{1}{2}(z+w) \sin \frac{1}{2}(w-z)$.

5.5.2. Find values of z such that $\cos z = 2$; and such that $\sin z = i$ (cf. § 5.8).

5.5.3. Show that if $w = \sin z$, then w is unbounded. (Hint: consider the exponential form, and consider $z = iR$, where R is real.)

5.5.4. Write the two square roots of $4 - 4i$ in modulus–amplitude form.

5.5.5. Write in $x + iy$ form the complex numbers given in modulus–amplitude form as (a) $6e^{2\pi i}$; (b) $e^{\pi i}$; (c) $e^{-\pi i}$; (d) $e^{i\pi/2}$; (e) $e^{-i\pi/2}$; (f) $\sqrt{(2)} e^{i\pi/4}$.

5.6.1. Express $1, -1, i, -i$ respectively in the form $\exp(i\theta + 2n\pi i)$, and hence express the four fourth roots of these expressions in modulus–amplitude form. In the case of the first two, reduce your answers to $x + iy$ form.

5.7.1. Obtain series for $\cosh z$, $\sinh z$.

5.7.2. Prove that (a) $\cosh^2 z - \sinh^2 z = 1$; (b) $\cosh^2 z + \sinh^2 z = \cosh 2z$.

5.7.3. Obtain some other hyperbolic relationships analogous to well-known trigonometrical results.

5.7.4. Show that $\sinh z$, $\cosh z$ are regular functions throughout the whole z-plane (see § 5.11).

5.7.5. Solve for z the equation $\tanh z = (e^z - 1)/(e^z + 1)$.

5.7.6. Solve the equation $\sinh z = 3$, giving your answer in the form of a logarithm (see § 5.8).

5.7.7. Prove that

(a) $\sinh(x + iy) = \sinh x \cos y + i \cosh x \sinh y$;

(b) if $\sin(x + iy) = \sinh(u + iv)$, then $\tan x \tan v = \tanh y \tanh u$.

5.7.8. If $u + iv = \tanh(x + iy)$, express u and v in terms of x and y.

5.8.1. Calculate: (a) $\text{Log } i$; (b) $\text{Log }(-1)$; (c) $\text{Log }(1 + i)$; (d) $\text{Log }(1 - i\sqrt{3})$.

5.8.2. From the equation $d/dz[\log(1 + z)] = 1/(1 + z)$ deduce a series for $\log(1 + z)$ in positive integral powers of z. Attempt to derive similar series for (a) $\log(2 + z)$, (b) $\log(i + 2z)$. Determine the circle of convergence for the series in each case.

5.8.3. Show that it is not in general true that $\log z + \log w = \log zw$ (where principal logarithms are used) and determine conditions under which it will be true. Give the more general statement which will always be true for addition of logarithms of complex numbers.

5.8.4. Determine at what points in the plane, if any, the following functions of z are not defined. (a) $\log(8 - z^3)$; (b) $\log(1 + z^2)$; (c) $\log \sinh z$; (d) $\log \cos z$.

5.8.5. Show that

(a) $\log z$ is not continuous at points on the negative part of the real axis;

(b) if z traces out a closed path which encircles the origin once, and $w(z)$ is a value of $\text{Log } z$ chosen to vary continuously with z, then $w(z)$ does not regain its original value when z does;

(c) if a cut be made along the negative part of the real axis, so that z cannot describe a closed path around the origin without crossing the cut, then $\log z$ will be regular in the remainder of the complex plane.

5.9.1. Show that Z^z (a) has only one value if and only if z is a real integer, (b) has a finite number of different values if and only if z is real and rational.

5.9.2. Prove that, if we take the principal value of the left-hand side, $(\cos \theta + i \sin \theta)^z = \cos z\theta + i \sin z\theta$. Comment on other possible values.

5.9.3. Find all the values of (a) i^i; (b) $(-1)^i$; (c) $(-1)^{-i}$; (d) e^{3+4i}; (e) $(1+i)^i$; (f) $(1 + i\sqrt{3})^{2i}$.

5.9.4. Determine whether or not the following results are true for complex values of the variables; consider principal and general values: (a) $Z^{-z} = 1/Z^z$; (b) $Z^z W^z = (ZW)^z$; (c) $(W^z)^Z = W^{zZ}$; (d) if $z^w = 1$, then either $z = 1$ or $w = 0$.

5.10.1. Assuming that

$$\tan^{-1} z = \sum_0^\infty a_r z^r,$$

and that a power series expansion of a regular function is unique, determine a_r ($r = 0, 1, 2, \ldots$) by term-by-term differentiation and comparison with another series.

5.10.2. Assuming that

$$f(z) = 1 + \sum_1^\infty w(w-1)(w-2)\ldots(w-r+1)z^r/r!,$$

where w is a complex constant, and that term-by-term differentiation is permissible, prove that $df(z)/dz = wf(z)/(1+z)$. Verify that $f(z) = (1+z)^w$ is a solution of this differential equation.

5.11.1. Justify the term-by-term differentiations in the previous two examples, stating limitations on $|z|$.

5.11.2. In the proof of the theorem of 5.11:

(a) prove preliminary result (1) in detail by both suggested methods;

(b) use Cauchy's method of multiplication of series in order to prove result (2);

(c) state at what point the proof would break down if we assumed that $r \leqslant \varrho$ instead of $r < \varrho$;

(d) state why it was necessary to introduce $\varrho < R$, and why it would not have been possible to omit ϱ completely and simply assume $r < R$, $r+m < R$?

(e) in the proof, we expanded the expression $(z+h)^n$, replaced z and h in the resulting expression by their moduli r and m, and, finally, reverted to the original form by rewriting the expression as $(r+m)^n$. Why did we not avoid all this by simply replacing z, h by r, m respectively in the original expression $(z+h)^n$?

5.11.3. A power series can be given in powers of $(z-c)$, where c is complex, instead of in powers of z. By considering a change of variable, examine the validity of the theorem of §5.11 for the more general power series, and explain the significance geometrically.

5.11.4. Devise complex series which:

(a) converge at all points on the circumference of the circle of convergence;
(b) converge at no point on the circle of convergence;
(c) converge at some points, but not all, on the circumference.

Determine a series which converges everywhere on the circumference of the circle of convergence, yet after two differentiations converges nowhere on this circumference.

5.11.5. Regarding integration as anti-differentiation, investigate term-by-term integration of a power series. [Write $f(z) = s_n(z) + R_n(z)$, where $s_n(z)$ is the sum of the first n terms of the series.]

Is it possible for a series to have, after integration term by term, a larger circle of convergence?

5.12.1. By assuming

$$\sinh z = \sum_0^\infty a_r z^r$$

and differentiating twice, determine the coefficients in the series for $\sinh z$. (It may be assumed that the series for $\sinh z$ is unique. NB: $\sinh 0 = 0$, $\cosh 0 = 1$.)

5.12.2. Adopt the successive differentiation technique in order to determine power series—assumed to exist and to be unique—for the following functions: (a) $\tanh^{-1} z$; (b) $\log(1-z)$; (c) a^z (a real). Determine the radius of convergence for each series.

5.13.1. Prove that:

(a) $\sin^{-1} z = -i \log [iz \pm \sqrt{(1-z^2)}]$,

(b) $\cosh^{-1} z = \pm \log [z + \sqrt{(z^2-1)}]$,

(c) $\tanh^{-1} z = \tfrac{1}{2} \log [(1+z)/(1-z)]$.

Examine whether the removal of the ambiguity of sign in (a) and (b) would impose any limitations on the domain of z or the range of the function, and identify any points at which any of the inverse functions, trigonometric and hyperbolic, are undefined.

CHAPTER 6

STRAIGHT LINE AND CIRCLE

6.1. The Standard Equation of a Straight Line

If a, b, and c are three real numbers, and if a, b are not both zero, then the set of points $z = x+iy$ whose real and imaginary parts satisfy the equation $ax+by+c = 0$ is a straight line. By means of the relations $z = x+iy$, $\bar{z} = x-iy$, we can rewrite the equation of this line in the form

$$a(z+\bar{z})/2+b(z-\bar{z})/2i+c = 0,$$

or, on simplifying,

$$z(a-ib)+\bar{z}(a+ib)+2c = 0.$$

If we now write $a-ib = A$, so that $a+ib = \bar{A}$, we may write the equation in the simpler form

$$Az+\bar{A}\bar{z}+B = 0,$$

where A, $\bar{A}$ are conjugate and B is real. Conversely, any equation of this form will represent a straight line provided that the complex number A is not zero.

Note that both z and $\bar{z}$ are involved in the equation. A linear equation involving z only would have precisely one solution, not a whole line of them.

6.2. Other Forms of the Equation

(a) (*Amplitude form*). Since the significant feature of a straight line is its constant direction, a simple form of equation to a straight line may be derived as follows: if the amplitude of the directed line segment

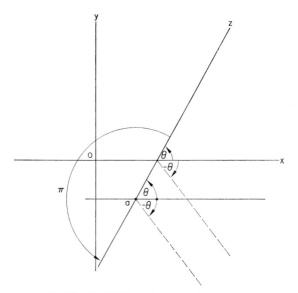

FIG. 6.1. The straight line. Amp $(z-a) = \theta$ or $\theta+\pi$.

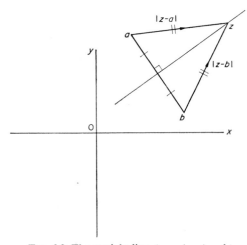

FIG. 6.2. The straight line. $|z-a| = |z-b|$.

joining a point a on the line to another point z of the line is θ, evidently we may write amp $(z-a) = \theta$, which, if a is fixed and z varies, gives us the equation to the "half-line" issuing from a in the direction θ. The other half of the line is then given by the equation amp $(z-a) = \pi+\theta$. (It is important to note that the equation to the second half-line is NOT amp $(z-a) = -\theta$, which gives a half-line making an angle 2θ with the first, and equally inclined to the real axis). (Fig. 6.1.)

(b) (*Modulus form*). Since any point on the perpendicular bisector (mediator) of the join of two points is equidistant from the points, and conversely, we may express the equation of a line l in the form $|z-a| = |z-b|$, where a and b are any two points equidistant from l such that the join of a and b is perpendicular to l (i.e. b is the reflection of a in l). (Fig. 6.2.)

6.3. The Circle

Obviously the equation of a circle with centre a and radius r is $|z-a| = r$. There are, however, several other useful forms of equation to a circle.

(a) The standard equation for a circle in the (x, y) plane is

$$x^2+y^2+2gx+2fy+c = 0,$$

where $f^2+g^2-c > 0$. If we transform this equation by means of the substitutions $x+iy = z$, $x-iy = \bar{z}$, we obtain

$$(z+\bar{z})^2/4+(z-\bar{z})^2/(-4)+g(z+\bar{z})+f(z \quad \bar{z})/i \mid c = 0,$$

which reduces to

$$z\bar{z}+Az+\bar{A}\bar{z}+c = 0,$$

where $A = g-if$, $\bar{A} = g+if$.

Conversely, it is easy to check that any equation of this form, with the coefficients of z and $\bar{z}$ conjugate and the constant term c real, and with $c < A\bar{A} = |A|^2$, represents a circle with centre $-\bar{A}$ and radius $\sqrt{(A\bar{A}-c)}$. (Exercise 6.3.1.)

(b) (*Amplitude form*). Since the angle subtended by an arc of a circle at any point on the remainder of the circumference is constant, we

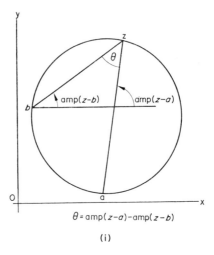

$$\theta = \text{amp}(z-a) - \text{amp}(z-b)$$

(i)

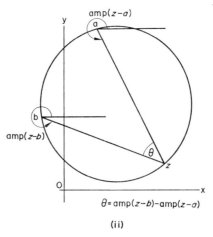

$$\theta = \text{amp}(z-b) - \text{amp}(z-a)$$

(ii)

FIG. 6.3. Arc of a circle. Amplitude form.

obtain the equation of an arc of a circle passing through two points a and b and containing an angle θ in the form

$$\text{amp}\,(z-a) - \text{amp}\,(z-b) = \theta, \quad \text{or} \quad -\theta,$$

according to the relative positions of a, b and the arc in question. (Fig. 6.3.)

Since division of complex numbers involves subtraction of their amplitudes, we can write instead

$$\text{amp } [z-a)/(z-b)] = \theta, \quad \text{or} \quad -\theta.$$

We shall naturally wish to know the equation for the "missing" arc in each case. However, since the opposite angles of a cyclic quadrilateral add up to π, it is easy to see from a diagram (Fig. 6.4) that the equation of the remainder of the circle of which one arc is

$$\text{amp } [(z-a)/(z-b)] = \theta, \quad \text{is} \quad \text{amp } [(z-a)/(z-b)] = \theta-\pi.$$

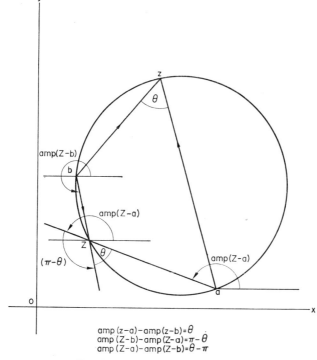

FIG. 6.4. The circle. Amplitude form.

(c) Let A and B be two fixed points in the complex plane, and let P be a variable point which moves so that $AP/PB = k$, where k is a constant $\neq 1$. Then the internal and external bisectors of APB, since they divide AB internally and externally respectively in the ratio AP/PB, meet the line through AB in two fixed points, X and Y say. The angle XPY is easily seen to be a right angle, whence the locus of P is a circle on AB as diameter. Hence, if the complex numbers corresponding to A, B and P are a, b and z respectively, then the equation $|(z-a)/(z-b)| = k, (k \neq 1)$, represents a circle (Fig. 6.5). As k varies, we obtain a family of circles with a common line of centres—the line

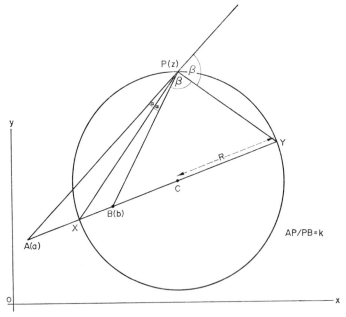

FIG. 6.5. The circle with equation $|(z-a)/(z-b)| = k \ (k > 1)$.

containing A, B. We shall later (Exercise 6.3.6) show that the family is a coaxal system.

We will examine further the relation of the points A, B to the circle on diameter XY. Let C be the mid-point of XY, i.e. the centre of the circle; then we may write the equation AX/XB in the form $(CX-CA)/(CB-CX) = (CY+CA)/(CY+CB)$. If we therefore let $CX = CY = R$ and cross-multiply, we obtain

$$(R-CA)(R+CB) = (R+CA)(CB-R),$$

which on expansion reduces simply to $CA \cdot CB = R^2$. This shows that A and B are inverse points (see § 7.7) with respect to the circle.

We now show that, conversely, if a and b represent two inverse points with respect to a circle, we may write the equation of that circle in the form $|(z-a)/(z-b)| = k$, with a suitably chosen k (Fig. 6.6).

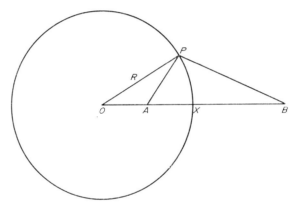

FIG. 6.6. A circle and two inverse points. $OA \cdot OB = OP^2$.

In the diagram, O is the centre of the circle with radius R, the points A and B are inverse with respect to the circle, and P is any point on the circumference. Since $OA \cdot OB = R^2 = OP^2$, it follows that $OA/OP = OP/OB$. Hence the triangles OAP and OPB, which have a common angle at O and sides about the angle proportional, are similar. It follows that $AP/PB = R/OB$, which is a constant (i.e. independent of the position of P). If we denote this constant by k, we have $|(z-a)/(z-b)| = k$ as the equation of the given circle. This is what we wished to prove.

(Evidently, if $k = 1$, A and B would coincide.)

It is worth considering what happens in the limiting case as R approaches infinity, and the circle becomes in the limit a straight line. If AB meets the circle at X, we have

$$(R-AX)(R+BX) = R^2 \quad \text{or} \quad R(BX-AX) = AX \cdot BX.$$

Thus $BX-AX = AX \cdot BX/R$, which approaches zero as R approaches infinity. So in the limit, $BX = AX$, while the line BXA is perpendicular to the straight line through X which represents the limiting position of the circle. In other words, two points which are mirror images of each other with respect to a straight line l may be regarded as inverse points with respect to l when l is considered as the limiting case of a circle whose radius becomes infinite.

Exercises

6.1.1. Derive the $(z, \bar{z})$ equations of the following lines:

(a) the line passing through the point ci (where c is real) and making an angle θ with the real axis, where $\tan \theta = m$;
(b) the line through the points a and ib (a, b real);
(c) the line through the points x_1+iy_1 and x_2+iy_2;
(d) the line through the points z_1, z_2;
(e) the line through the point x_1+iy_1 making an angle $\tan^{-1} m$ with the real axis;
(f) the real axis;
(g) the imaginary axis;
(h) lines parallel to the axis;
(i) a line through 0 making an angle $\tan^{-1} m$ with the real axis;
(j) a line such that the perpendicular to the line from the origin is of length p, and makes an angle θ with the real axis.

6.1.2. Determine the condition that the line $az+\bar{a}\bar{z}+c = 0$ is (a) parallel, and (b) perpendicular to the line $bz+\bar{b}\bar{z}+d = 0$.

6.1.3. Find the equation of a line through $p+iq$ (p, q real) (a) parallel to, and (b) perpendicular of the line $az+\bar{a}\bar{z} + c = 0$.

6.1.4. Find the equation of the perpendicular bisector of the line joining $a+ib$ and $c+id$.

6.1.5. Determine what is represented by each of the following equations:

(a) $z^2+6iz-9 = 0$;
(b) $z^2+a^2 = 0$ (a real);
(c) $az+b = 0$ (a, b complex).

6.2.1. Give the equation of the line through the point $1+i$ and making an angle $\pi/4$ with the positive real axis. Determine the angle made with real axis by the line joining $1+i$ to the origin. (Use the amplitude form.)

6.2.2. Rewrite answers to Exercise 6.1.1 (a)–(i) in amplitude form.

6.2.3. Rewrite answers to Exercise 6.1.1 (f), (g), (h) and (j) using the modulus form. (Hint: for (j) find the reflection of the origin in the line.)

6.2.4. Determine the complex number representing Z if Z is the reflection in $az+\bar{a}\bar{z}+c = 0$ of the point $p+iq$. Note in particular the reflection of $p+iq$ in (a) the real axis, and (b) the imaginary axis. If $p+iq = w$, express these results in terms of w, $\bar{w}$.

6.3.1. Verify the statement of the last paragraph of § 6.3(a).

6.3.2. Express the equation to the unit circle (with centre 0 and radius 1) in at least six different forms. (Hint: note, for example, that i, $-i$ are ends of a diameter; that $\frac{1}{2}$, 2 are inverse points—among many other pairs).

6.3.3. Give in amplitude form the equation to that semicircle on the line segment joining 0 to $2i$ as diameter, which lies in the second quadrant. Give also the equation to the corresponding semicircle lying in the first quadrant.

6.3.4. State in amplitude form the equation to that part of the unit circle which lies below the real axis. Verify from your equation that the point $\frac{1}{2}-\frac{1}{2}i\sqrt{3}$ lies on this semicircle.

6.3.5. Obtain, in any appropriate form, the equation to each of the following circles:

(a) the circle with $a+ib$, $p+iq$ as ends of a diameter;
(b) the circle with centre $2+3i$, and passing through the origin;
(c) the circle through the points 0, 1, i;
(d) the circle through the points z_1, z_2, z_3. (Give an analytical condition under which this circle will degenerate into a straight line.)
(e) a circle having z_1 and z_2 as inverse points and passing through the point z_3.

6.3.6. Devise further examples as in Exercise 6.3.5.

6.3.7. Devise a test to determine whether the points z_1, z_2, z_3 and z_4 are concyclic. [Note Exercise 6.3.5(d).]

6.3.8. In Fig. 6.5, let M be the mid-point of AB. Let A, B be fixed and $k (= AP/PB)$ vary. Show that the points X and Y will not remain fixed, but prove that $MX \cdot XY$ remains constant ($= MB^2$). Deduce that if $r^2 = MX \cdot MY$, then a circle with centre M and radius r will intersect all the circles of the system § 6.3(c) orthogonally.

Prove, analytically or geometrically, that the system is a system of coaxal circles with limiting points A and B, and with radical axis the perpendicular bisector of AB.

CHAPTER 7

SIMPLE TRANSFORMATIONS

7.1. Translation

Let $a = r_0 e^{i\theta_0}$ be a given complex number, and let us consider the mapping from the complex plane to itself defined by taking the point z to the point $z+a$. We may indicate such a mapping by means of either of the two forms $z \to z+a$ or $z' = z+a$, according to convenience. In the latter case, z' is, of course, regarded as lying in the same complex plane as z. If S is any set of points in the plane, then the image of S under this mapping has the same geometrical configuration as S; more precisely, the transformation affects neither distances between points nor angles between lines. All that has happened is that every point has been moved a distance r_0 along a line making an angle θ_0 with the real axis. Such a mapping, in which every point moves the same distance in the same direction, is known as a *translation*. If a is a real number, the translation is parallel to the real axis; and if it is a pure imaginary number, the translation is in the direction of the imaginary axis.

It is easily seen that the repeated application of translations can be reduced to one translation. (See Exercise 7.1.1.)

A translation may be considered as moving the frame of reference, without rotation, magnification, or shrinking.

7.2. Reflection

If any point z in the complex plane is mapped upon its conjugate $\bar{z}$ (i.e. $z \to \bar{z}$ or $z' = \bar{z}$), the effect is as though every point has been replaced by its reflection in the real axis. Reflection in the imaginary axis can similarly be effected by means of the mapping $z \to -\bar{z}$; for this clearly takes $z = a+ib$ to $-(a-ib) = -a+ib$. Another simple example is the mapping $z' = i\bar{z}$, which produces reflection in the line $x = y$. Since only points on the line of reflection reflect into themselves, we may deduce a z, $\bar{z}$ form of the equation to the line $x = y$ as being $z = i\bar{z}$.

It is evident that the result of two reflections cannot be that of a single reflection, since the sense of angles is reversed by one reflection, but not by an even number of reflections.

7.3. Rotation

If the complex plane is rigidly rotated about the origin through an angle θ, relative to the original axes, the point $z = re^{i\phi}$ is transformed to z', where $z' = re^{i(\theta+\phi)} = e^{i\theta}z$. Thus the transformation may be expressed in the form $z' \to e^{i\theta}z$ or $z' = cz$, where c is the complex number $e^{i\theta}$. Conversely, the mapping $z \to cz$, where $|c| = 1$, corresponds to a rotation of the plane about the origin through an angle θ, where $\theta = \text{amp } c$. The result of a sequence of rotations is again a rotation, and neither distances between points nor angles between lines are altered by rotations of the plane.

7.4. Magnification

It is natural next to consider the effect of multiplication by a number not of unit modulus. We consider first the case of multiplication by a real number $k > 0$. The transformation $z' = kz$ will clearly have no effect on amplitude, but all distances from the origin will be altered in the ratio $k : 1$. This has the effect of a magnification of magnitude k, with centre 0. (The term *magnification* is still used when $k < 1$. If $k = 1$ we regard the transformation as an identity magnification.) Therefore it follows that the transformation $z' = ke^{i\theta}z$ is a combination of, first, a rotation through θ, and then a magnification k. In fact the two processes are easily seen to commute, so that the order of the

124

operations is not significant. Since any complex number c can be expressed in the form $ke^{i\theta}$, where $k = |c|$ and $\theta = $ amp c, we have now shown that the effect of the transformation $z' = cz$ is that of a rotation combined with a magnification.

7.5. Glide Reflection

If we combine a reflection with a translation along the axis of reflection, as, for instance, in the mapping $z' = \bar{z}+a$ (a real), the resulting transformation is called a glide reflection.

7.6. Shear

In all the transformations so far considered, or in any combination of them, all geometrical figures will retain their shape (specifically angles and ratios of corresponding lengths are unaltered). Their size may be affected (by magnification). In the transformation now to be considered, shape is not in general retained. The effect of the transformation $z' = z+k\mathcal{J}(z)$ (k real), which is described as a *shear* parallel to the real axis, is to convert a rectangle with sides parallel to the axes into a parallelogram (Fig. 7.1). A shear will not affect areas (consider figures divided into thin strips parallel to the real axis, in the above case) but will in general affect angles.

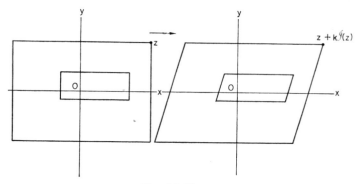

FIG. 7.1. Shear.

7.7. Inversion

Consider the mapping $z \to 1/z$, defined everywhere except at the origin. In modulus–amplitude form this may be written $re^{i\theta} \to (1/r)e^{-i\theta}$. Thus every point except the origin is mapped to a point whose distance from the origin is the reciprocal of the original distance and whose amplitude is the negative of its former value. If we now follow this transformation by a reflection in the real axis, which preserves modulus while changing the sign of the amplitude, we have the transformation $re^{i\theta} \to (1/r)e^{i\theta}$. This transforms every point z to another one on the same radius vector through the origin, such that the product of the two distances from the origin is 1. We observe that points on the unit circle are invariant under this transformation, points outside it are transformed to points in the interior, and interior points are transformed to exterior points. Since $e^{i\theta} = 1/e^{-i\theta}$, this transformation may be expressed in the form $re^{i\theta} \to 1/re^{-i\theta}$ or $z' = 1/\bar{z}$, and it is known as *inversion* with respect to the unit circle (Fig. 7.2).

Geometrically, inversion with respect to a circle with centre C and radius k is defined as the transformation which maps every point P on to a point P' such that (i) $CP \cdot CP' = k^2$, and (ii) CPP' is a straight

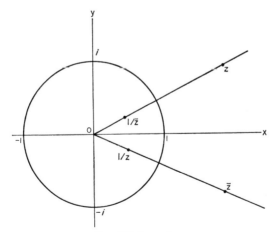

Fig. 7.2. Inversion.

line. (In the case we have considered, C is 0 and $k = 1$.) The following results are easily proved either geometrically or analytically (see Exercise 7.7.1):

(a) A straight line through C, the centre of inversion, inverts into itself.
(b) A straight line not passing through C inverts into a circle.
(c) A circle passing through C inverts into a straight line.
(d) A circle not passing through C inverts into a circle.

Other results relating to inversion are to be found in the exercises and in Chapter 8. The fact with which we are here concerned is that the transformation $w = 1/z$ is a combination of inversion and reflection in the real axis from which its effect on straight lines and circles may easily be deduced.

7.8. The Point at Infinity

In applying the process of inversion with respect to the unit circle, we note that the inverse of 0 remains undefined. It is not even determined as regards direction, since the origin lies on all the radii of the circle. Conversely there is no point outside the circle which has the origin as its inverse. To preserve the one-to-one correspondence between point and inverse which otherwise exists, we add to the system of complex numbers in the complex plane a single "point at infinity", which we regard as the inverse of the centre of the circle of inversion, and hence also as the point which inverts into that centre. We write $1/z = \infty$ when $z (= \bar{z}) = 0$, and refer to the plane with infinity appended as the *extended complex plane*. We shall find that in practice no difficulties arise about the failure to attach an amplitude to the point at infinity, although a point moving along any straight line in the complex plane "approaches infinity" in a specific direction.

By the use of the point at infinity we can unify the four theorems listed in § 7.7. We may simply regard a straight line as a circle which passes through the point at infinity, and then all four theorems are covered by the single statement that "the inverse of a circle is a circle".

It should be noted that we use the phrase "z approaches infinity" to indicate that $|z|$ increases without limit.

Exercises

7.1.1. Prove that:

(a) the combination of two translations is a translation (the "identity translation" $z' = z$ is admissible);

(b) the combination of translations is associative; $[T_1 + (T_2 + T_3) = (T_1 + T_2) + T_3]$;

(c) every translation T has an inverse $(-T)$ such that $T + (-T) = I$, the identity translation;

(d) the combination of translations is commutative; $(T_1 + T_2 = T_2 + T_1)$;

(e) lengths are unaltered by translation;

(f) the direction of lines is unaltered by translation;

(g) if a translation keeps one point invariant, then it is the identity translation;

(h) given a translation T there is a translation T' such that the coordinates of $T(z)$ with respect to the original axes $0x$, $0y$ are the same as the coordinates of z with respect to the translated axes $T'(0x)$, $T'(0y)$. Identify the translation T' in relation to T.

7.1.2. Find a translation which will reduce the equation

$$z\bar{z} + az + \bar{a}\bar{z} + c = 0 \quad (c < a\bar{a})$$

to the form $z'\bar{z}' = r^2$, and interpret geometrically.

7.2.1. The points of the plane are reflected in the line $az + \bar{a}\bar{z} + c = 0$. Identify:

(a) all invariant points;

(b) all invariant lines.

Determine the position of the image of $p + iq$:

(c) in the case $a = 1$;

(d) in the case $c = 0$;

(e) in the general case.

7.2.2. Show that reflection leaves unchanged (a) distances between points, and (b) angles between lines.

Show by means of a counter-example that the combination of reflections is not commutative.

7.3.1. Show that:

(a) any two rotations about the origin when combined are equivalent to a single rotation;

(b) that the combination of rotations is associative;

(c) that there is an identity rotation;

(d) that to each rotation there is an inverse rotation which combined with it will give the identity;

(e) that the combination of rotations is commutative;

(f) that lengths are unaltered by rotations;

(g) that the angle between two lines is unaffected by a rotation;

(h) that the combination of a rotation about the origin and a reflection in a line through the origin can be reduced to a reflection in some other line.

Investigate whether the operations described in (h) are commutative.

7.4.1. Investigate magnification as Exercise 7.3.1, showing that properties corresponding to (a), (b), (c), (d), (e), and (g) hold.

7.4.2. Consider the effects of various combinations of the transformations above (translation, reflection, rotation, magnification) on straight lines, angles, triangles, squares, and circles. Is the set of transformations so far considered "closed" under combination of transformations? (i.e. can any combination of two always be replaced by a single transformation of one of the types considered?)

7.5.1. Find the transformation formula for a glide reflection in the line $z = i\bar{z}$.

7.5.2. Find the formula for a glide reflection consisting of a reflection in the line $az + \bar{a}\bar{z} + c = 0$ followed by a glide of distance d parallel to the line.

7.5.3. Examine the combination of two glide reflections. Consider whether the set of translations, rotations, reflections, and glide reflections is closed under combination of transformations.

Now include magnification and consider the new situation.

7.5.4. Show that any glide reflection may be regarded as the result of three ordinary reflections in different lines, not in general concurrent.

7.6.1. Examine whether the set of shears with a common invariant point and a common axis is closed under combination of shears. Is the combination of shears (a) associative, (b) commutative? Is there an identity shear? Has each shear an inverse? Determine any invariant points and lines under shear.

Area is preserved under shear. Can you find any other geometrical properties which are preserved? (For example, mid-points of lines? Ratio of division of lines by a point? Parallelism? Bisectors of angles? Lengths of segments?)

7.7.1. Prove that:

(a) a straight line through the centre of inversion inverts into itself;

(b) any other straight line inverts into a circle;

(c) a circle through the centre of inversion inverts into a straight line;

(d) any other circle inverts into a circle.

(Use either geometrical or algebraic methods; preferably both in each case.) Determine all the invariant points and invariant lines of the transformation.

7.7.2. Show that a set of lines through a point inverts into a set of coaxal circles; and that a set of coaxal circles inverts in general into another such set. Indicate the special case.

Show that the inverse of the inverse is the original configuration.

7.7.3. Show geometrically that a circle orthogonal to the circle of inversion inverts into itself. (Hint: consider the inverses of the tangents to the given circle from the centre of inversion, and the inverses of the points of contact.)

7.7.4. Find the inverse with respect to the unit circle of the parabola whose cartesian equation is $y^2 = 4ax$. Investigate some other curves and their inverses. (Try graphical plotting of inverses.)

7.8.1. Show that each of the four theorems of § 7.7 can be regarded as a "circle" theorem by means of the correct use of the idea of a "point at infinity".

7.8.2. Determine on to which point of the complex plane the point at infinity is mapped by means of the following mappings; determine also which points, if any, are mapped on to the point at infinity: (a) $z' = e^{-z}$; (b) $z' = az+b$; (c) $z' = (az+b)/(cz+d)$; (d) $z' = 1/(z^2+1)$.

CHAPTER 8

CONFORMAL TRANSFORMATIONS

8.1. Regular Transformations

Let f be a regular function in a domain D of the z-plane and let $w = f(z)$. We may regard f as the transformation $z \to w$. For convenience we shall regard the w-plane as distinct from the z-plane; then to every point in D there corresponds a point in the w-plane. To the whole of D will correspond a set D' in the w-plane.

Consider two curves C_1 and C_2 in D, which intersect at the point z and each have a definite (unique) tangent at that point. Since the transformation is continuous, their images in the w-plane will be two curves K_1, K_2, intersecting at the corresponding point w. We now consider two small displacements of z; one, $\delta_1 z$, along C_1, and the other, $\delta_2 z$, along C_2. Again, by continuity, to these will correspond small displacements along K_1, K_2. Let $\delta_1 z = re^{i\theta_1}$, $\delta_2 z = re^{i\theta_2}$, where r is small, and let the corresponding displacements of w be $\delta_1 w = R_1 e^{i\phi_1}$ $\delta_2 w = R_2 e^{i\phi_2}$ respectively. Then R_1 and R_2 will also be small; but the same does not, of course, apply to the various amplitudes, which will approximate to the angles the tangents at z or w to the appropriate curves make with the corresponding real axis. (We here assume that K_1 and K_2 *have* tangents at w. It is not difficult to show that if $f'(z) \neq 0$, then this is always the case. See Exercise 8.1.1.)

The situation is illustrated in Fig. 8.1.

Now since f is regular, we may write

$$\delta_1 w / \delta_1 z = f'(z) + u_1(\delta_1 z), \quad \text{where} \quad u_1(\delta_1 z) \to 0 \quad \text{as} \quad r = |\delta_1 z| \to 0,$$
$$\delta_2 w / \delta_2 z = f'(z) + u_2(\delta_2 z), \quad \text{where} \quad u_2(\delta_2 z) \to 0 \quad \text{as} \quad r = |\delta_2 z| \to 0.$$

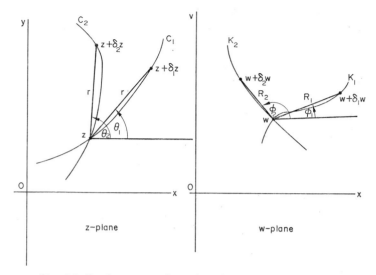

FIG. 8.1. Continuous transformation of two intersecting curves.

Hence

$$\delta_1 w/\delta_1 z = \delta_2 w/\delta_2 z + \varepsilon, \qquad \text{where} \quad \varepsilon \ [= u_1(\delta_1 z) - u_2(\delta_2 z)]$$

approaches zero as r approaches zero.

Thus we have

$$(R_1/r)e^{i(\phi_1-\theta_1)} = (R_2/r)e^{i(\phi_2-\theta_2)} + \varepsilon,$$

which may be arranged in the form

$$(R_1/R_2)\exp i(\phi_1-\phi_2) = \exp i(\theta_1-\theta_2) + \varepsilon \cdot (r/R_2)\exp i(\theta_1-\phi_2).$$

Providing, therefore, that $\lim_{r\to0} (r/R_2)$ is finite (see below for further discussion of this point), it follows, on writing $\lim_{r\to0} (\phi_1-\phi_2) = \beta$, $\lim_{r\to0} (\theta_1-\theta_2) = \alpha$, and noting the vanishing of ε as r approaches zero, that

$$\lim_{r\to0} (R_1/R_2)\exp i\beta = \exp i\alpha.$$

Now two complex numbers which are equal must have the same modulus and amplitude, whence $\alpha = \beta$ and $R_1/R_2 \to 1$. Geometrically this means that the angle between the tangents to the transformed curves is equal to the angle between the tangents to the original curves, both angles being measured in the same sense; while the "local magnification" R/r is the same in all directions (since $R_1/r \to R_2/r \to |f'(z)|$ as r decreases towards zero). This implies that any small figure in the z-plane is transformed into an approximately similar figure in the w-plane. In general the magnification factor will vary from point to point.

DEFINITION. *A mapping which preserves angles between curves meeting at z is said to be* conformal *at z. A mapping which is conformal at each point z in a domain D is said to be* conformal *in D.*

(It is to be noticed that the sign as well as the magnitude of the angles is to be preserved; put another way, the sense of rotation from one tangent to the other is the same for K_1 to K_2 as for C_1 to C_2.)

We return to the remark "provided that $\lim_{r\to 0} (r/R_2)$ is finite". If it should be infinite the proof, of course, breaks down (and in fact the preservation of angles does not usually then occur). The implication in this case is that $\lim_{r\to 0} (R_2/r) = 0$, which is equivalent to $f'(z) = 0$. This case will be examined later. Meanwhile we exclude points where the derivative is zero from our considerations. (See Exercise 8.1.3 for an investigation of this matter.)

8.2. The Bilinear Transformation

We have already considered in Chapter 7 several conformal transformations. (It should, however, be noted that inversion, $w = 1/\bar{z}$, is not conformal, since the sense of angles is not preserved.) We will now examine in greater detail the transformation

$$w = (az+b)/(cz+d),$$

where a, b, c, and d are complex numbers, and we assume for the moment that (i) $c \neq 0$, (ii) a and b are not both zero, (iii) $ad \neq bc$. [Of course, (iii) will imply (ii).] The point of stipulation (iii) is that if $a/c = b/d = k$, the equation reduces to $w = k$ unless $z = -b/a$ $(= -d/c)$, and w is indeterminate in this case. Thus the transformation is not one-to-one and certainly not conformal.

Solving for z we easily obtain

$$z = (dw - b)/(-cw + a).$$

This inverse transformation and the original one together show that there is a one-to-one correspondence between points in the z-plane and points in the w-plane induced by the bilinear transformation—apart from two important exceptions. The point $z = -d/c$ has no (finite) point corresponding to it, and is said to transform to the "point at infinity" in the w-plane; whilst no point in the z-plane will transform to the point $w = a/c$, which is similarly regarded as the transform of the "point at infinity" in the z-plane. [It is to be noticed that as $|z|$ increases without limit, w approaches the value a/c, as may be seen by writing $w = (a + b/z)/(c + d/z)$.]

With this understanding we may regard the transformation as completely one-to-one in the extended complex plane. It is known as the bilinear, or Möbius, transformation.

8.3. Straight Lines and Circles

As we saw in § 6.3, the equation $|(z - p)/(z - q)| = k$ represents a straight line or circle according as k is or is not 1. In either case, the points p and q may be regarded as inverse. On applying the bilinear transformation we readily see that the equation of the transform in the w-plane is

$$|\{(dw - b)/(-cw + a) - p\}/\{(dw - b)/[(-cw + a) - q]\}| = k,$$

which simplifies to

$$|\{(d + pc)w - (ap + b)\}/\{(d + qc)w - (aq + b)\}| = k$$

or

$$|(d + pc)/(d + qc)| \, |\{w - (ap + b)/(cp + d)\}/\{w - (aq + b)/(cq + d)\}| = k,$$

which we may write

$$|(w-P)/(w-Q)| = K,$$

where $$P = (ap+b)/(cp+d),$$

$$Q = (aq+p)/(cq+d), \quad K = k|(cq+d)/(cp+d)|.$$

The equation to the transform is hence seen to be either that of a circle with P and Q as inverse points, or, if $K = 1$, that of a straight line, again with P and Q inverse with respect to it. We note also that P and Q are the transforms of p and q respectively. We hence have a theorem.

THEOREM. *A bilinear transformation maps a circle and two inverse points into a circle and two inverse points, where we agree to treat a straight line as a circle passing through the point at infinity.*

It is less difficult than the above calculations would seem to imply, to determine whether or not a given circle transforms into a straight line. We need merely to check whether or not the circle passes through the point which will transform to the point at infinity; if and only if it does so, will the transform be a straight line. Also, straight lines through this point will evidently remain straight lines after transformation.

8.4. The Mapping of a Domain

Suppose we have a simple closed curve C in the z-plane; this will divide the points of the plane into three sets—interior points (i.e. points inside C), exterior points (points outside C), and points on C. (The fact that this is true for *any* simple closed curve is a difficult theorem— the Jordan curve theorem—but for the curves that we shall usually consider, consisting of arcs of circles and straight line segments, it will be easy to see that it is true.) On applying a conformal transformation f, which takes no point of C to infinity, the curve C maps into a simple closed curve $f(C)$ in the w-plane. However, an interior point of C in the z-plane may map into either an interior or an exterior point of $f(C)$ in the w-plane. (For example, consider C as the unit circle, and the transformations $w = z$ and $w = 1/z$.) We have the following theorem.

THEOREM. *If a function f is regular and one-to-one in a domain D, and D contains within it a simple closed curve C, which is mapped by f into a simple closed curve f(C), then either all interior points of C map into the interior of f(C), or all map into the exterior of f(C), and similarly for exterior points of C.*

Proof. Suppose that P and Q are two interior points of C in the z-plane. Then they may be joined by a curve which has no points in common with C. Let them be transformed by f to points X and Y respectively in the w-plane, and suppose that X is an interior point of $f(C)$, and Y is an exterior point of $f(C)$. Then by the continuity of the transformation the curve joining P and Q will be transformed into a continuous curve joining X and Y, and thus necessarily having a point in common with $f(C)$. Such a common point must have been the transform of a point of C. However, the curve joining P, Q had no point in common with C. This is a contradiction to the assumption that f was one-to-one, and hence proves that our supposition about X and Y was false; so they must either both be interior points or both be exterior points. This proves the theorem.

The theorem can be extended to the case when, for example, C is a straight line and f is a bilinear transformation, to show that the whole of one side of C is transformed to the whole of one side of $f(C)$ when $f(C)$ is a straight line, or to the whole interior or the whole exterior of $f(C)$ when $f(C)$ is a circle. Similar statements can be made when, say, C is a simple closed curve, but $f(C)$ is not (C containing the point transformed to infinity).

We illustrate various ways of investigating a transformation in the following example.

EXAMPLE. Determine the transform of the upper half-plane under the mapping $z \rightarrow w$, where $w = (z+i)/(z-i)$.

Method 1. We require first the transform of the boundary of the given domain, viz. the real axis $y = 0$, whose equation can be written $(z-\bar{z})/2i = 0$ or $z = \bar{z}$. From the relation $z = (iw+i)/(w-1)$ we obtain the equation to the transform of the real axis in the form

$$i(w+1)/(w-1) = -i(\bar{w}+1)/(\bar{w}-1),$$

which simplifies to

$$w\bar{w}-w+\bar{w}-1+w\bar{w}+w-\bar{w}-1 = 0,$$

i.e. $w\bar{w} = 1$, which is the unit circle $|w| = 1$.

The point i in the z-plane transforms to the point at infinity in the w-plane—a point external to the unit circle. Thus the upper half z-plane transforms to the exterior of the unit circle in the w-plane.

Method 2. Consider the transforms of the points 1, 0, -1 on the boundary line $y = 0$. These points transform respectively to $(1+i)/(1-i)$, -1, $(-1+i)/(-1-i)$, or, on reduction to the $a+ib$ form, to i, -1, $-i$. The only circle through these points is the unit circle $|w| = 1$. The rest of the investigation proceeds as before.

Method 3. Since i and $-i$ are reflections of each other in the real axis, and hence inverse points with respect to this line, the equation of the axis may be written in the form

$$|(z-i)/(z+i)| = 1.$$

(There are, of course, an infinite number of such pairs of inverse points available.)

On substituting $z = (iw+i)/(w-1)$ we easily obtain

$$|(iw+i-iw+i)/(iw+i+iw-i)| = 1,$$

which reduces to $|w| = 1$.

Method 4. Expressing the equation of the real axis in the amplitude form amp $z = 0, \pi$ (for the two half-lines from the origin) and substituting for z as before gives

$$\text{amp } \{i(w+1)/(w-1)\} = 0, \pi.$$

Now amp $i = \pi/2$, and multiplication involves addition of amplitudes; hence we easily obtain the equation

$$\text{amp } \{(w+1)/(w-1)\} = \pm\pi/2,$$

which represents a circle on the join of -1, 1 as diameter—again the unit circle $|w| = 1$.

Method 5. Transforming to cartesian coordinates gives us the relation, on substituting $w = u+iv$, $z = x+iy$,

$$x+iy = i[(u+1)+iv]/[(u-1)+iv].$$

From this we easily obtain

$$y = (u^2-1+v^2)/[(u-1)^2 + v^2],$$

whence, since the denominator is real and non-negative, $y > 0$ implies

$$u^2 + v^2 > 1,$$

whence, once again, we find the required transform to be the exterior of the unit circle.

Thus there is a variety of methods available for attempting to solve problems relating to bilinear (and other) transformations. Often the methods above do not all work equally well, and part of the skill involved lies in selecting the line of attack most appropriate to the particular situation.

8.5. Half-plane to Interior of Circle

We will examine the problem of finding the most general bilinear transformation $w = (az+b)/(cz+d)$ which will transform the half-plane $\mathcal{R}(z) > 0$ on to the interior of the unit circle $|w| = 1$, and the imaginary axis into this circle.

We first make use of the fact that "inverse points transform into inverse points", and of the observation that 0 and infinity are inverse points of the unit circle. These points in the w-plane are the transforms of $-b/a$ and $-d/c$ respectively in the z-plane, so that $-b/a$ and $-d/c$ must be reflections of each other in the imaginary axis. Figure 8.2 shows that if $-b/a = p$, then we must have $-d/c = -\bar{p}$, and we may therefore write our transformation in the simpler form

$$w = (a/c)(z-p)/(z+\bar{p}),$$

where a/c may be treated as a single complex number.

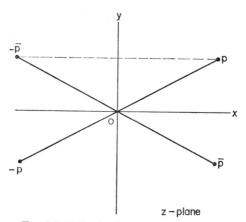

FIG. 8.2. Reflection in the imaginary axis.

We now look for a point on the imaginary axis which it will be easy to transform, use the fact that its transform must lie on the unit circle, and attempt thereby to obtain some further limitation on the constants in the transformation formula. The origin is obviously the simplest point to consider. Letting $z = 0$ gives $|a/c| \, | -p/\bar{p}| = 1$, which evidently reduces to $|a/c| = 1$. Hence we may replace a/c by $e^{i\alpha}$, say.

The transformation has now been reduced to the form

$$w = e^{i\alpha}(z-p)/(z+\bar{p}),$$

and, with two "degrees of freedom" available (we may still select α and p to suit our convenience), we have to ensure that any point on the imaginary axis transforms to a point on the circumference of the unit circle. As it happens, however, no further steps are necessary to ensure this.

Figure 8.3 indicates that any point on the imaginary axis is equidistant from p and $-\bar{p}$, so that

$$|(iy-p)/(iy+\bar{p})| = 1.$$

Hence $|w| = 1$ if $z = iy$ (where y is real), and we have transformed the boundary of the half-plane into the boundary of the unit disc.

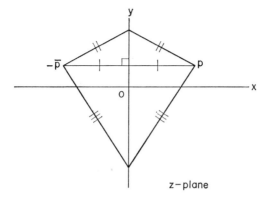

FIG. 8.3. Points on the imaginary axis are equidistant from p and $-\bar{p}$.

It still remains to ensure that points in the half-plane $\mathcal{R}(z) > 0$ transform into the interior of the circle. But since $z = p$ transforms to the interior point $w = 0$, the additional requirement is very simple —merely that $\mathcal{R}(p) > 0$. The problem is therefore now solved and the solution is

$$w = e^{i\alpha}(z-p)/(z+\bar{p}),$$

where α is real and $\mathcal{R}(p) > 0$.

8.6. Circle to Circle

We now consider the problem of finding all the bilinear transformations which transform the unit circle $|z| = 1$ into the unit circle $|w| = 1$. Again we start with the observation that 0 and the point at infinity in the w-plane have arisen from inverse points (say p and $1/\bar{p}$ respectively) with respect to the circle $|z| = 1$. This enables us to reduce $w = (az+b)/(cz+d)$ to the form

$$w = (a/c)(z-p)/(z-1/\bar{p}) \quad \text{or} \quad w = (a\bar{p}/c)(z-p)/(\bar{p}z-1).$$

We next select a point on $|z| = 1$ which is easily dealt with, and require that the transform shall lie on $|w| = 1$. The origin is not this time suitable, but $z = 1$ is the obvious choice for reasons of simplicity. On substitution, our requirement leads to

$$|a\bar{p}/c|\,|(1-p)/(\bar{p}-1)| = 1,$$

whence, since $|1-p| = |1-\bar{p}|$, we find $|a\bar{p}/c| = 1$, or $a\bar{p}/c = e^{i\alpha}$ say (α real).

The transformation now reads

$$w = e^{i\alpha}(z-p)/(\bar{p}z - 1).$$

To investigate what further, if anything, needs to be done we substitute a general point on the unit z-circle, say $z = e^{i\theta}$. This gives

$$w = e^{i\alpha}(e^{i\theta}-p)/(\bar{p}e^{i\theta} - 1),$$

which may be written

$$w = e^{i(\alpha-\theta)}\,(e^{i\theta}-p)/(\bar{p}-e^{-i\theta}).$$

Since the conjugate of $e^{i\theta}-p$ is $e^{-i\theta}-\bar{p}$, and since conjugate complex numbers have the same modulus, it follows immediately that $|w| = 1$ for all points w which are transforms of points z with $|z| = 1$. The problem is thus already solved.

If we wish the interiors to correspond, it is easily seen that p, which transforms to the origin, must lie inside the unit z-circle, i.e. $|p| < 1$. So that the most general bilinear transformation which transforms the unit circle and its interior in the z-plane to the unit circle and its interior in the w-plane is given by

$$w = e^{i\alpha}(z-p)/(\bar{p}z - 1),$$

where α is real and $|p| < 1$.

8.7. Other Simple Transformations

We now proceed to consider transformations other than the bilinear ones. In order to determine the general effect of a mapping we will try to see what it does to some simple configurations, such as lines parallel to the coordinate axes, or lines through the origin. What will specially interest us is which type of curve will transform into a straight line parallel to one of the axes, since it may well be that by such a transformation we can reduce some given geometrical problem to a simpler form. For instance, we might be able to change a family of curves into a set of parallel lines and thus easily find their orthogonal trajectories; on transforming back we should have the family of orthogonal trajectories of the original system, since the angles at which curves intersect are unaffected by conformal transformations.

A difficulty of which we have to be aware is that our transformations need not be one-to-one over the entire plane; so that we need to pay special attention to which domains, if any, are transformed in a one-to-one way.

8.8. The Mapping $w = z^2$

Writing this in the form $u+iv = (x+iy)^2$ and expanding, we easily obtain $u = x^2-y^2$, $v = 2xy$, from which we see at once that the lines $u = $ const., $v = $ const. in the w-plane are the transforms of two families of hyperbolas in the z-plane, namely $x^2-y^2 = $ const. and $xy = $ const. These families cut each other orthogonally, as is well known and can easily be established using analytical geometry. It follows here, though, from the fact that the transformation is conformal (at least when $z \neq 0$) since $f'(z) = 2z \neq 0$ away from the origin, and the transforms of the curves of different families are clearly orthogonal (in this case, straight lines). (Fig. 8.4.)

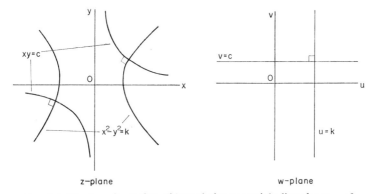

FIG. 8.4. The transformation of hyperbolas to straight lines by $w = z^2$.

To determine the transform of the line $x = c$ (a real constant) we eliminate the variable y from the equations

$$u = c^2 - y^2, \qquad v = 2cy,$$

to obtain

$$u = c^2 - v^2/4c^2 \quad \text{or} \quad v^2 = 4c^2(c^2 - u),$$

a family of parabolas. A similar calculation with regard to $y = k$ gives us

$$v^2 = 4k^2(u + k^2),$$

which represents another family of parabolas. Since the transformation is conformal away from the origin, we know that the second family must be orthogonal to the first, and it is not difficult to verify directly by analytical geometry that this is in fact so.

Let us also consider the transformations in the w-plane of the loci $r = \text{const.}$, $\theta = \text{const.}$ respectively in the z-plane; i.e. the circles with centre the origin, and the half-lines radiating from the origin.

From $z = re^{i\theta}$ we obtain at once $w = r^2 e^{2i\theta}$, giving

$$u = r^2 \cos 2\theta, \qquad v = r^2 \sin 2\theta.$$

Hence if r is constant, we find on eliminating θ that $u^2 + v^2 = r^4$, showing that the family of concentric circles maps into the same set of concentric circles in the w-plane; yet that only the unit circle maps into itself—circles outside the unit circle map into larger ones, those inside into smaller ones.

Since $w = r^2 e^{2i\theta}$ we see that if $\theta = \text{const.}$ then amp $w = \text{const.} = 2\theta$ to within a multiple of 2π. Therefore the half-line obtained by rotating the positive x-axis through an angle θ is transformed into the half-line obtained by a rotation of 2θ. Note, however, that the *principal* amplitude of $f(z)$ is not necessarily twice the *principal* amplitude of z; for example, the principal amplitude of $(-1+i)$ is $3\pi/4$, but the principal amplitude of $(-1+i)^2 = -2i$ is $-\pi/2$ and not $6\pi/4 = 3\pi/2$.

Nevertheless, we see that if we transform two half-lines radiating from the origin by means of the mapping $w = z^2$, we obtain two other such lines which now have twice as large an angle between them. This, of course, seems not to agree with the previous theoretical result that angles are unchanged by a regular transformation. The reason, though, is simply that at the origin $f'(z) = 0$, and, as mentioned previously, at such points (critical points) the theorem does not apply.

Another point is worth noting. As $z = e^{i\theta}$ moves round the unit circle counter-clockwise from $+1$ to -1, the corresponding point w will move, also counter-clockwise, all the way round the circle from $+1$ back to $+1$ again. If z continues to progress along the lower semicircle, w will start to trace out the unit circle for the second time, and one-to-one correspondence will have been lost. One way of restoring this correspondence is to consider the w-plane to be cut along the positive part $(u \geqslant 0)$ of the real axis and not to allow w to move along any curve which crosses the cut. This corresponds to restricting z either to the upper half-plane $y > 0$ or to the lower half-plane $y < 0$. We then see that the upper half-z-plane maps on to the whole of the w-plane, minus the cut, in a one-to-one way, and, similarly, for the lower half-z-plane. In a similar way we could cut the w-plane along some other half-line radiating from the origin, instead of the u-axis, and this would likewise decompose the z-plane into two regions on each of which f would be one-to-one.

8.9. The Exponential Transformation

The mapping $w - e^z$, or $u+iv = e^{x+iy}$, immediately leads to the relations $u = e^x \cos y$, $v = e^x \sin y$. The lines $x = c$, where c is a constant become circles $u^2+v^2 = e^{2c}$ (> 0), and the lines $y = k$ (const.) transform into the lines $v/u = \tan k$. These radial lines are, of course, orthogonal to the concentric circles $u^2+v^2 = $ const. The curves in the z-plane which transform into the lines $u=$const., $v=$const. are obviously $e^x = u \sec y$, $e^x = v \operatorname{cosec} y$, respectively.

Looking again at the transforms of the lines $y = $ const. we easily see that the mapping is by no means one-to-one. For if we replace y by $y+2\pi$, we find that for any x the point $x+iy+2\pi i$ gives the same u, v, and hence the same w, as does $x+iy$. It is easily seen that the strip of the z-plane given by $-\pi < y \leqslant \pi$, $-\infty < x < \infty$ maps on to the whole of the w-plane except for the origin (which corresponds to $x = -\infty$). Any strip given by

$$-\pi + 2n\pi < y \leqslant \pi + 2n\pi \quad (n = \pm 1, \pm 2, \ldots)$$

will similarly transform on to the whole of the w-plane. Narrower strips of the same kind transform into wedges with vertices at $w = 0$, the point $w = 0$ itself being excluded. (Fig. 8.5.)

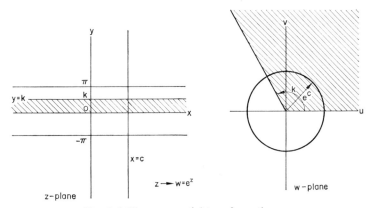

FIG. 8.5. The exponential transformation.

8.10. The Logarithmic Transformation

The transformation $w = \text{Log } z$ is not a genuine (one-valued) function (see § 5.8) since there are an infinite number of values of w corresponding to any non-zero z. Let us initially examine $w = \log z$, meaning that we concern ourselves with the principal value of the logarithm.

The relation

$$u + iv = \log \sqrt{(x^2 + y^2)} + i \tan^{-1} y/x$$

(more precisely i amp z) shows that the lines $u = \text{const.}$, $v = \text{const.}$, are the transforms of circles $x^2 + y^2 = e^{2u}$ and rays (half-lines) $y = vx$. This is only to be expected, since this transformation is the inverse of the one we have just studied, $w = e^z$. In fact, because of this inverse relationship we can derive any results we need about log z by considering e^z instead. So we need not discuss this transformation further except to note briefly that by limiting ourselves to principal values we limited ourselves to a strip $(-\pi < v \leqslant \pi)$ of the w-plane. Similar strips of the form $(2n-1)\pi < v \leqslant (2n+1)\pi$ correspond to other (non-principal) values of the logarithm.

8.11. The Transformation $w = k \sin z$

We will examine briefly this trigonometrical transformation. The relation

$$u+iv = k(\sin x \cosh y + i \cos x \sinh y),$$

which we obtain on expanding $\sin (x+iy)$, leads to

$$u = k \sin x \cosh y,$$
$$v = k \cos x \sinh y.$$

These equations give the curves which transform to $u = $ const., $v = $ const. respectively as

$$\sin x \cosh y = \text{const.},$$
$$\cos x \sinh y = \text{const.}$$

To find the transform of the line $x = c$, we eliminate y from the equations

$$u = k \sin c \cosh y,$$
$$v = k \cos c \sinh y,$$

to obtain

$$u^2/k^2 \sin^2 c - v^2/k^2 \cos^2 c = 1,$$

which is the equation of a hyperbola; or, as c varies, a family of confocal hyperbolas. Similarly for the line $y = b$, eliminating x gives

$$u^2/k^2 \cosh^2 b + v^2/k^2 \sinh^2 b = 1,$$

which, as b varies, gives a family of confocal ellipses orthogonal to the hyperbolas; in fact, as may easily be shown, these hyperbolas and ellipses all belong to the same confocal family of conics. This family represents the transform of the sets of lines parallel to the axes in the z-plane.

Since the sin and cos of $x+2\pi$ are respectively equal to $\sin x$ and $\cos x$, it is clear that to see the effect of the transformation $w = k \sin z$ on the whole z-plane we need consider only a vertical strip of width 2π—this will map on to the complete w-plane (check this).

8.12. The Transformation $w = z+c^2/z$ (c real)

We first determine the critical points of this transformation, i.e. those points where $dw/dz = 0$. Since $dw/dz = 1-c^2/z^2$ these points are given by $z = \pm c$. (We assume $c > 0$ for convenience.) We notice in passing that dw/dz approaches 1 as z approaches infinity, from which we easily deduce that local magnification is approximately unity at a great distance from the origin.

The substitution $z = x+iy$ is in this example less informative than the use of the modulus–amplitude form $z = re^{i\theta}$. We easily obtain

$$w = u+iv = re^{i\theta}+(c^2/r)e^{-i\theta}$$

giving immediately

$$u = (r+c^2/r) \cos \theta, \quad v = (r-c^2/r) \sin \theta,$$

which show that if r is constant and θ varies, i.e. if z moves on a circle with centre the origin and radius r, then the locus of w is an ellipse $u^2/a^2+v^2/b^2 = 1$, where $a = r+c^2/r$, $b = r-c^2/r$. We will assume that $r > c$, which implies that the circle contains the critical points in its interior. Now if we replace r by c^2/r, which would then be less than c in this case, we obtain the same ellipse as before. Thus in general two circles transform to the same ellipse. It can in fact be shown (see Exercises) that the interior and the exterior of the circle $|z| = c$ in the z-plane each transform into the entire w-plane except for the straight line segment $[-2c, 2c]$. The transform of the circle $|z| = c$ is evidently this segment. (For, if $r = c$, then $u = 2c \cos \theta$, $v = 0$, $0 \leqslant \theta < 2\pi$.)

The inverse transformation

$$z = \tfrac{1}{2}[w + \sqrt{(w^2 - 4c^2)}]$$

is useful in transforming ellipses into circles. There is ambiguity, of course, in relation to the choice of square root. However, by a suitable choice we can ensure that when w approaches infinity so does z, so that the exteriors of the circle and ellipse correspond.

Exercises

8.1.1. In the notation of § 8.1, and letting $f'(z) = Re^{i\beta}$, deduce, by writing the left-hand side of the equation

$$\delta_1 w/\delta_1 z = Re^{i\beta} + u_1(\delta_1 z)$$

in modulus amplitude form that if $R \neq 0$, then the curve K_1 has a definite tangent at w.

8.1.2. Invent some non-trivial transformations (a) in which magnification is 1 everywhere; and (b) in which magnification is 2 everywhere. Attempt to express them algebraically and also to interpret them geometrically.

8.1.3. Let $f(z)$ be expressible in powers of $(z - z_0)$ in the form $f(z) = f(z_0) + a(z - z_0)^{n+1} + b(z - z_0)^{n+2} + \ldots$ $(a \neq 0)$.

(a) Show that each term in the expansion of $f'(z)$ has a factor $(z - z_0)^n$ and that $f'(z_0) = 0$. (We say that $f'(z)$ has a zero of order n at z_0.)

(b) Show that $f(z_1) - f(z_0) = a(z_1 - z_0)^{n+1} + \ldots$.

(c) If $z_1 - z_0 = \delta_1 z = re^{i\theta_1}$ and $f(z_1) - f(z_0) = \delta_1 w = R_1 e^{i\phi_1}$, show that $R_1 e^{i\phi_1} = Be^{i\alpha} r^{n+1} e^{i(n+1)\theta_1} + \ldots$ (terms of higher degree in r), where $Be^{i\alpha} = a$, in modulus–amplitude form.

(d) Deduce that $\lim_{r \to 0} \phi_1 = \alpha + (n+1) \lim_{r \to 0} \theta_1$.

(e) Write down the corresponding result for another displacement $\delta_2 z = re^{i\theta_2}$ inducing $\delta_2 w = R_2 e^{i\phi_2}$.

(f) Determine the angle between the tangents at w_0 to the curves K_1 and K_2, which are the images of the curves C_1 and C_2, along which z_1 and z_2 respectively move towards z_0, giving your result β in terms of the angle λ between the tangents to C_1 and C_2 at z_0.

(g) Determine whether or not it is possible for λ and β to be equal at any point where $f'(z) = 0$.

8.2.1. Examine the bilinear transformation for the case $c = 0$, $d \neq 0$. Express the nature of the transformation geometrically.

Examine also the result of taking other combinations of a, b, c, and d to be zero (including all of them).

8.2.2. *Critical points* of a function f are points where $f'(z) = 0$ or infinity. Determine the critical points of a bilinear transformation and of its inverse. Show that in any case where the constants a, b, c, and d are such that the transformation is not one-to-one, then neither is the inverse transformation.

Regarding the z and w-planes as the same, determine in terms of a, b, c, and d any points which are invariant under the transformation.

8.2.3. Consider the following mappings:

$$z_1 = cz, \quad z_2 = z_1 + d, \quad z_3 = 1/z_2, \quad z_4 = (cb/a - d)z_3, \quad z_5 = 1 + z_4, \quad w = (a/c)z_5,$$

where a, b, c, and d are non-zero complex numbers and $bc \neq ad$. State the effect of each transformation on straight lines and circles, and deduce that the set of all straight lines and circles in the z-plane is transformed into the set of all straight lines and circles in the w-plane by the transformation $w = (az + b)/(cz + d)$.

8.3.1. By direct use of the z, $\bar{z}$ forms of the equations to straight lines and circles, deduce that the bilinear transformation maps them into circles, or straight lines, and determine the condition that straight lines are obtained. Compare your condition with that given in § 8.3.

8.3.2. Express the unit circle $|z| = 1$ in a form which uses the pair of inverse points $\frac{1}{2}$, 2. Transform it by means of the mapping $w = (z - i)/(z + i)$ and compare your result with the one obtained by direct transformation of $|z| = 1$.

8.3.3. Express the real and imaginary axes in the form

$$|(z - a)/(z - b)| = 1,$$

and hence determine their images under the mapping $w = (z - 2i)/(2z + i)$.

Carry out the same transformation using z, $\bar{z}$ forms of the equations to the axes

8.3.4. Under the transformation

$$w = [(1 + i)z + 1 - i]/[(1 - i)z + 1 + i]$$

determine:

(a) the circle with centre the origin which transforms into a straight line;
(b) the condition that a circle or straight line in the z-plane will transform into a straight line;
(c) the common point through which pass all circles and straight lines in the w-plane which come from straight lines in the z-plane.

8.4.1. Examine the transforms of:

(a) the upper half-plane;
(b) the half-plane $\mathcal{R}(z) > 0$;
(c) the unit disc $|z| < 1$;
(d) the line $y = x$ $(z = i\bar{z})$;
(e) the circle amp $(z - 2)/z = \pm\pi/2$;

under the transformations:

(i) $w = (z+1)/(z-1)$;

(ii) $w = z/(z-2)$;

(iii) $w = (2z+i)/(z+2i)$;

(iv) $w = (cz+1)/(z+c)$ (c real, $c \neq 0, \pm 1$).

Use as many different methods as you can for each case. Also, regarding the z- and w-planes as the same, determine any invariant points for each of the transformations.

8.4.2. Examine whether there are lines which are invariant under non-trivial bilinear transformations. (A line L is *invariant* under f if and only if the image of L under f is L itself.)

8.4.3. Determine the image in the w-plane of that part of the z-plane given by $x > 1$, $y > 2$ under the transformation $w = (z+i)/(z-2)$.

8.5.1. Determine the most general bilinear transformation which will map the upper half-plane on to the unit disc $|w| < 1$ and the real axis on to the unit circle.

8.5.2. Determine the most general bilinear transformation which will map the upper half-plane on to itself.

8.5.3. Determine the most general bilinear transformation which will map the lower half-plane on to the interior of the circle $|w| = R$.

8.6.1. Determine the most general bilinear transformation which maps the interior of the circle $|z| = r$ on to the exterior of the circle $|w| = R$ (and the first circle into the second).

8.6.2. Find a bilinear transformation which will map the interior of the unit circle on to the interior of the circle $|w-i| = 1$ so that the points $z = 1$, $z = -1$, map into $w = 0$, $w = 2i$ respectively. Examine whether or not the solution you obtain is unique.

8.6.3. Examine the problem of mapping the interior of the unit circle onto a half-plane by means of a bilinear transformation.

8.6.4. Devise and try to solve some other simple bilinear transformation problems of the types above.

8.7.1. Determine what happens to:

(a) straight lines through the origin;

(b) circles with centre the origin, in the z-plane, under the mapping $w = z^n$ (n a positive integer).

(Hint: convert to modulus–amplitude form.)
Examine the conservation or otherwise of angles between lines passing through the origin, and comment.

8.7.2. Find a simple transformation which will map that part of the first quadrant between the line $x = y$ and the imaginary axis, including the lines themselves, on to the entire complex plane. (The mapping need not be one-to-one.) State what points and lines, if any, are invariant under the transformation.

8.7.3. Determine a transformation which will map the entire z-plane excluding the positive real axis upon the half-plane $\mathscr{R}(z) > 0$; examine some characteristics such as invariant points and lines, and the images of lines parallel to the coordinate axes. (Hint: modulus–amplitude form might help.)

8.8.1. Under the transformation $w = z^2$, determine the images of:

(a) the interior of the upper semicircle of the circle $|z-1| = 1$;

(b) the region between the curves $xy = a$, $xy = b$, $x^2 - y^2 = c$, $x^2 - y^2 = d$ (a, b, c, d real);

(c) the unit square with vertices $0, 1, 1+i, i$;

(d) the part of the half-plane $\mathscr{R}(z) < 0$ which lies outside the circle $|z| = 2$.

8.8.2. If $z^{1/2}$ is defined the number with modulus $\sqrt{|z|}$ and amplitude half the principal amplitude of z, find what type of region in the z-plane would transform under the mapping $w = z^{1/2}$ into a strip with edges parallel to the imaginary axis. Do the same for strips of the w-plane parallel to the real axis.

8.8.3. Find the invariant points, if any, of the transformation $z \rightarrow 1/2z^2 + \frac{1}{2}$. By reducing the transformation to a sequence of simpler transformations, or otherwise, examine some other geometric effects.

8.9.1. Determine the image in the w-plane of the region $\mathscr{R}(z) < 0$, $-\pi < \mathscr{I}(z) < 0$ under the transformation $w = \exp z$.

8.9.2. Examine the transformation $w = \exp mz$ (m real) for various m.

8.9.3. Determine the image of the square with vertices $0, 1, 1+i, i$ under the mapping $w = \exp z$. Consider also a similarly situated square of side 2π.

8.10.1. Investigate the mappings $w = \log z$, $w = \log(1+z)$, determining in each case the effect of the transformation on lines parallel to the axes, radial lines, circles with centre 0. (Note that the transformation $w = \log z$ is not continuous on the negative real axis, and not defined at the origin.)

8.11.1. Examine as in the text the transformations

$$w = \sinh z, \quad w = \cosh z, \quad w = \cos z, \quad w = \tan z, \quad w = \tanh z, \quad w = \sin^{-1} z, \text{ etc.}$$

8.11.2. By considering the application of two successive transformations, or otherwise, show that curves in the z-plane with equations of form $\sin x \cosh y = $ const. map, under the transformation $w = \sin^2 z$, into parabolas. Investigate other combinations of transformations.

8.12.1. Obtain the following results for the transformation $w = z + c^2/z$:

(a) conjugate points $(z, \bar{z})$ are mapped into conjugate points $(w, \bar{w})$;

(b) two points which are reflections of each other with respect to the imaginary axis in the z-plane, map into reflections with respect to the imaginary axis in the w-plane;

(c) the circle $|z| = c$ maps into the line segment joining the points $2c$, $-2c$. (Examine the transform of both the upper and lower semicircle; also each quadrant.) State what happens to the interior of $|z| = c$ under this transformation.

Investigate the transform of any circle passing through the points c, $-c$. [Hint: the circle may be written in amplitude form, amp $(z-c)/(z+c) = \alpha$, $\alpha - \pi$. Show that $w - 2c = (z-c)^2/z$, $w + 2c = $, etc.]

8.12.2. By actual plotting of a number of points, investigate the curve (a Joukowski aerofoil) obtained by transforming a circle which passes through the point -1 and contains the point 1 in its interior, and which lies in the z-plane, by means of the transformation $w = z + 1/z$. Prove that the exteriors of the two curves correspond.

By regarding the transformation $w = z + c^2/z$ as a combination of three successive transformations

$$z_1 = z/c, \quad z_2 = z_1 + 1/z_1, \quad w = cz_2,$$

discuss the effect of transforming the circles in the z-plane which pass through $-c$, and contain c in their interior, by means of the transformation $w = z + c^2/z$.

8.12.3. Determine any critical points of the transformation $w = z - c^2/z$, and the possibility of obtaining from it a Joukowski aerofoil. (Hint: consider a circle through $-ic$ and containing ic.)

8.12.4. Determine the curves in the z-plane which, under the transformation $w = z + 1/z$, give rise to the circle $|w| = 2$. [Hint: $(w-2)/(w+2) = (z-1)^2/(z+1)^2$.] Sketch the curves.

CHAPTER 9

INTEGRATION

9.1. Integration as Anti-differentiation

Let $f(z)$ and $\phi(z)$ be two functions defined in some domain D of the complex plane, and suppose that $f(z)$ is differentiable and that $f'(z) = \phi(z)$ everywhere in D. Then in relation to D we can write purely symbolically $\int \phi(z)\, dz = f(z)$ to describe this situation where we do not assume at this stage that for complex functions $\int$ has any other meaning. Of course, if the above holds for $f(z)$ and $\phi(z)$, then it also holds for $f(z)+c$ and $\phi(z)$, where c is any real or complex constant: therefore we will frequently write $\int \phi(z)\, dz = f(z)+c$ to take account of this. We shall say that we *integrate $\phi(z)$ with respect to z to obtain $f(z)+c$.*

According to this definition of $\int$ we can write down immediately such results as

$$\int z^n dz = z^{n+1}/(n+1)+c \quad (n \neq -1),$$

$$\int \exp z\, dz = \exp z + c,$$

$$\int \cos z\, dz = \sin z + c, \quad \text{etc.},$$

which hold everywhere in the complex plane (excluding the origin in the first equation if $n < -1$).

It is not at this stage clear whether there is any connection between the integral as defined above and the usual integral for functions of a real variable (defined by a limit of a summation process). We leave consideration of this question until later.

9.2. Integration as Summation

The definition of integration, for real functions of real variables, as the limit of a summation plays a most important part in the practical applications of the process (although a link with "anti-differentiation" frequently provides a short cut to the result). In attempting to apply the sum–limit approach in the field of complex variables, we meet with a difficulty not experienced in the real field. We recall that for real functions, $\int_a^b f(x)\, dx$ is the limit of a sum of areas of rectangles whose bases are on the real axis between a and b. In complex variable, on the other hand, the expression $\int_a^b$, where a and b may well be complex numbers, gives no indication as to the path along which z progresses from a to b. This choice of path will naturally be expected to affect the value of any summation, and evidently needs to be taken into account.

All the paths we shall consider will be regarded as being made up of a finite number of continuous segments which also satisfy a further condition. If the equation for such a segment is written parametrically in the form $x = x(t)$, $y = y(t)$, where the segment is described as t varies from t_1 to t_2, then we shall require that dz/dt be continuous (or, what is the same thing, that dx/dt and dy/dt both be continuous) on the closed interval $[t_1, t_2]$. This will ensure that dx/dt, dy/dt are integrable. Paths made up of a number of such segments will be called *contours*.

We now define *the integral of $F(z)$ along a segment S* from $z = z_1$ to $z = z_2$ [where $z_1 = z(t_1)$, $z_2 = z(t_2)$] as

$$\int_S F(z)\, dz = \int_{t_1}^{t_2} F[z(t)]\, (dz/dt)\, dt$$

(where F is sufficiently well-behaved for the integral on the right-hand side to exist), or, splitting the terms up into real and imaginary parts,

$$\int_{t=t_1}^{t=t_2} (u+iv)(dx+idy),$$

where
$$F(z) = U(x, y) + iV(x, y)$$
$$= u(t) + iv(t),$$
$$dx = (dx/dt)\, dt, \qquad dy = (dy/dt)\, dt$$

This integral may now be arranged as the sum of two terms in the form

$$\int_S F(z)\, dz = \int_{t_1}^{t_2} (u\, dx - v\, dy) + i \int_{t_1}^{t_2} (v\, dx + u\, dy),$$

where the u, v, dx, and dy are all expressed in terms of the parameter t. Since all the integrals are now real, they may be evaluated in the ordinary way.

Then if the contour C is made up of a sequence of segments $S_1 \ldots S_k$, we define *the integral of $F(z)$ along the contour C* as

$$\int_C F(z)\, dz = \int_{S_1} F(z)\, dz + \ldots + \int_{S_k} F(z)\, dz.$$

Thus we have defined the *contour integral* of $F(z)$ along C using essentially only the concept of integration of a real function over an interval. It is, however, to be emphasized that we must not at this stage assume that there is any connection whatsoever between this $\int_C F(z)\, dz$ and the $\int F(z)\, dz$ as described in § 9.1. The procedure will be illustrated by an example, which will also show the dependence of the result upon the path of integration. We shall take $F(z) = x - iy$, and integrate this expression, first, round the circumference C of the unit circle $|z| = 1$, starting and finishing at the point $z = 1$. Let this contour integral be denoted by I.

The obvious parameter to use is $\theta = \text{amp } z$, which varies from 0 to 2π. We then have $x = \cos\theta$, $y = \sin\theta$, $dx = -\sin\theta\, d\theta$, $dy = \cos\theta\, d\theta$, and

$$I = \int_0^{2\pi} (u\, dx - v\, dy) + i(v\, dx + u\, dy),$$

where $u = x$, $v = -y$, giving

$$I = \int_0^{2\pi} \cos\theta(-\sin\theta)\,d\theta - (-\sin\theta)\cos\theta\,d\theta$$

$$+ i\int_0^{2\pi} -\sin\theta(-\sin\theta)\,d\theta + \cos\theta\cos\theta\,d\theta$$

$$= \int_0^{2\pi} 0\cdot d\theta + i\int_0^{2\pi} 1\cdot d\theta$$

$$= i\cdot 2\pi \quad \text{or} \quad 2\pi i.$$

We will now consider the integral I' of the same function, evaluated round the square C' with vertices 1, $1+i$, i, and 0, where we proceed as before in a counter-clockwise direction, starting and finishing at $z = 1$.

Writing the integral in the form

$$\int_{C'} u\,dx - v\,dy + i(v\,dx - u\,dy)$$

$$= \int_{C'} x\,dx + y\,dy + i(-y\,dx - x\,dy),$$

we note that for the first segment S_1 (the line joining 1 to $1+i$) only y varies, and may be taken as the variable ($x = 1$ throughout); for the second segment S_2, $y = 1$, and x varies from 1 to 0, and so on. From the definition

$$\int_C = \int_{S_1} + \int_{S_2} + \int_{S_3} + \int_{S_4},$$

and so we have to evaluate each stage of the integral separately. For the real parts, we find that $\int_{C'} x\,dx + y\,dy$ reduces to

$$\int_0^1 y\,dy + \int_1^0 x\,dx + \int_1^0 y\,dy + \int_0^1 x\,dx$$

which evidently sums to zero. For the imaginary part of the integral, we have to evaluate

$$i\left(\int_0^1 1\cdot dy + \int_1^0 -1\cdot dx + \int_1^0 0\cdot dy + \int_0^1 -0\cdot dx\right),$$

the other parts vanishing because either dx or dy is 0 throughout the segment considered. The resulting integral I' is found to be $2i$, which we see differs from the previous result for I even although beginning and end points were the same for C' as for C.

In the above example it so happened that C had the same point for its beginning and end (and similarly for C'). For an example where this is not the case, it may be noted that the integral from 0 to 1 of $x - iy$ directly along the real axis is $\frac{1}{2}$, while along the other three sides of the square C', from 0 to i, then to $1 + i$, and from there to 1, the result is $\frac{1}{2} - 2i$, as may be seen from the above calculations.

9.3. Integration as Summation. Another Approach

Let us recall the definition of the integral of a real function f over an interval $[a, b]$ of the real line. We divide $[a, b]$ into a number of smaller intervals by introducing points $x_0, x_1, \ldots, x_n$ with $a = x_0 < x_1 < x_2 < \ldots < x_n = b$, and then compute the sum

$$S = \sum_{r=0}^{n-1} f(\xi_r)(x_{r+1} - x_r),$$

where ξ_r is any point chosen on the interval $[x_r, x_{r+1}]$. As we introduce more points x_r so that the maximum length of the sub-intervals $[x_r, x_{r+1}]$ approaches zero it may happen that S approaches some finite limit L. If this is so, and if the limit L does not depend on the choice of the ξ_r, then we say that f is *integrable* over $[a, b]$ and that its *integral* is L; and we write

$$\int_a^b f(x)\, dx = L.$$

Now let f denote a function of a complex variable, and let C be a continuous path along which f is defined, joining the complex number a to the complex number b. We saw in § 9.2 that (if C is sufficiently "smooth") we can define a contour integral of f along C by taking a parametrization of C and evaluating some real integrals.

However, we can also imitate directly the above definition of integral for a real function as follows: we take points $z_0, z_1, \ldots, z_n$ along C such that $z_0 = a$, $z_n = b$, and z_{r+1} lies on the path from z_r to b $(0 \leqslant r \leqslant n-1)$, and choose points Z_r $(0 \leqslant r \leqslant n-1)$ such that Z_r lies on the path z_r to z_{r+1} of C. Let S now denote

$$\sum_{r=0}^{n-1} f(Z_r)(z_{r+1} - z_r)$$

which is, of course, a complex number. As before, as the number of points z_r on C is increased and the paths $z_r z_{r+1}$ become smaller, the sum S may approach a (complex) limit L. If so, and if L does not depend on the choice of the Z_r, then we say that f is *integrable* along C, and that its *integral* along C *is* L.

It can be proved, as in the case of real variable, that if f is continuous on C (or even if it is continuous on C except at a finite number of points), then f is integrable along C.

We thus have now two definitions of the integral of f along C: the limit of a sum as above, and the contour integral (defined using t) as in § 9.2. However, it can be shown quite easily that when these integrals both exist they are the same. From now on we shall call them both the (contour) integral of f along C, and denote them by $\int_C f(z)dz$.

To see how the limit–sum definition can apply in practice we will consider two examples.

(a) Let us first consider the case $f(z) = c$, where c is a constant, which may be complex. Then the sum S evidently reduces to

$$S = c[(z_1 - z_0) + (z_2 - z_1) + \ldots + (z_n - z_{n-1})] = c(z_n - z_0) \quad \text{or} \quad c(b-a).$$

There are two interesting features of the above result. One is that the integral has, surprisingly, been found without our having to specify the path from a to b, and thus must be independent of the particular path selected. The other point to note is that if we had tried to treat the problem of calculating

$$\int_a^b f(z)\, dz = \int_a^b c\, dz$$

by the procedure of anti-differentiation (which is what we would have done in the case of a real variable), we would have obtained formally $[cz]_a^b = c(b-a)$ again. We shall consider both these points again in relation to our next example.

(b) We take for our second example the "next simplest" function, $f(z) = z$. Since in this case z will vary as we move along the contour C, we have a choice of sums S to consider (for we can choose Z_r arbitrarily on $z_r z_{r+1}$).

Two possibilities are

$$S_1 = z_0(z_1-z_0)+z_1(z_2-z_1)+\ldots+z_{n-1}(z_n-z_{n-1})$$

or

$$S_2 = z_1(z_1-z_0)+z_2(z_2-z_1)+\ldots+z_n(z_n-z_{n-1}),$$

corresponding to the choices $Z_r = z_{r-1}$, $Z_r = z_r$ respectively.

Neither of these sums, as it happens, simplifies. However, since they tend to the same limit (because z is continuous so its integral exists) this limit will also be the limit of $S' = \frac{1}{2}(S_1+S_2)$, their mean. And on examining this we notice that most of the terms of S_1 and S_2, after expansion and addition, cancel out, leaving us with the result

$$S' = \frac{1}{2}(z_n^2-z_0^2) = \frac{1}{2}(b^2-a^2).$$

Again we have no need to proceed further to find the limit we require, since S' is constant. And again we observe that, firstly, the result is independent of the path, and, secondly, that had we proceeded as though faced with a problem in real variable, and applied anti-differentiation, we would have obtained the correct result:

$$\int_a^b z \, dz = (\tfrac{1}{2}z^2)_a^b = \tfrac{1}{2}(b^2-a^2).$$

From these examples and the examples of § 9.2 we therefore see that in some cases the integral depends on the path of integration from a to b and in others it does not. We will consider an extremely important theorem (Cauchy's theorem) relating to this in the next chapter. In the meantime we consider the relationship which seems to exist between anti-differentiation and sum–limit integration in the particular case when the function to be integrated is the derivative of a differentiable function.

9.4. Sum–limit and Anti-differentiation

THEOREM. *Let $F(z)$ be regular in a domain D and have derivative $F'(z) = f(z)$. Let C be a contour in D joining two points $z = a$, $z = b$. Then*

$$\int_C f(z)\, dz = F(b) - F(a).$$

Proof. Let us regard the contour as expressed in terms of a parameter t, which is such that when $t = t_1$, $z = a$, and when $t = t_2$, $z = b$. We shall assume also that dz/dt exists at all points of the contour. (If the contour consists of a number of segments for which this is true, but it is not true for the end points of the segments, the theorem is proved by considering one segment at a time and piecing together the results.)

The integral may then be written in the form

$$I = \int_{t_1}^{t_2} f[z(t)]\, \frac{dz}{dt}\, dt,$$

or, since $dF/dz = f(z)$,

$$I = \int_{t_1}^{t_2} \left(\frac{dF}{dz}\right) \frac{dz}{dt}\, dt,$$

where the expression in brackets would be converted into a function of the real variable t before integration proceeded.

However, just as with real variable we have $dF/dz \cdot dz/dt = dF/dt$, and

$$\int_{t_1}^{t_2} \frac{dF}{dt}\, dt = [F\{z(t)\}]_{t_1}^{t_2} = F(b) - F(a).$$

This proves the theorem; and thus links up the two integration procedures of §§ 9.1 and 9.2 in the cases where § 9.1 can be applied. We shall occasionally use the term "direct integration" in the exercises for the procedure of anti-differentiation followed by the insertion of limits of integration.

9.5. An Important Inequality

THEOREM. *If C is a contour of length L, and if $|F(z)| \leqslant M$ for all z lying on C, then $\left| \int_C F(z)\, dz \right| \leqslant ML$.*

Proof. Since the modulus of a sum is less than or equal to the sum of the moduli we may write, considering the integral as the limit of a sum,

$$|S| \leqslant \sum_0^{n-1} |f(Z_r)|\, |z_{r+1}-z_r| \leqslant M \sum_0^{n-1} |z_{r+1}-z_r|,$$

and the right-hand term can be made arbitrarily close to ML by suitable choice of the z_r. Hence the modulus of the required integral, which is the limit of $|S|$ as $|z_{r+1}-z_r|_{\max}$ approaches zero, is also less than or equal to ML.

This theorem is of great use in showing that certain parts of a contour integral may contribute negligibly to the total; we shall use it often in subsequent chapters.

Exercises

9.1.1. Integrate with respect to z, i.e. find a function whose derivative is:

(a) az^2+bz+c $(a, b, c$ complex constants);

(b) ze^{z^2};

(c) ze^{2z};

(d) $1/z$;

(e) $(3z+2)/(z-1)$;

(f) $\cos 5z$;

naming in each case any points where the original function or the derivative is undefined.

9.2.1. Integrate the functions defined by the following expressions:

(a) round the unit circle counter-clockwise from $z = 1$;

(b) round the square with vertices $1, 1+i, i, 0$ (counter-clockwise from 1 to 1);

(c) along the parabola $y = x^2$ from $z = 0$ to $z = 1+i$.

(i) $x+iy$; (ii) x^2-y^2+2ixy; (iii) x^2-y^2-2ixy; (iv) $x+y$; (v) $2x+3iy$, (vi) x^n $(n > 0)$; (vii) k (a complex constant).

9.3.1. By using the fact that as z_r approaches z_{r+1} the three expressions z_r^2, z_{r+1}^2, and $z_r z_{r+1}$ approach the same limit, construct a plausible argument to suggest that the integral, by direct summation, of z^2 along a curve joining the points $z = a$, $z = b$ is $(b^3-a^3)/3$. Indicate where your argument fails to provide a complete proof.

9.3.2. Experiment as in the previous question with the integral of (a) z^3 and (b) z^n (n a positive integer) along a suitable curve from $z = a$ to $z = b$. Comment on your results when $a = b$.

9.4.1. If C is any contour connecting the points $z = a$, $z = b$, calculate directly the integral with respect to z, along C, of the following functions:

(a) $e^z (\cos y + i \sin y)$;

(b) $\sin(pz+q)$ (p, q complex);

(c) $(3z+2)/(z-1)$ (the contour in this case must not pass through the point $z = 1$. Why?)

(d) $1/z^2$. (What restriction is there on the contour in this case?)

9.4.2. Evaluate the integral with respect to z of the function $1/z$ round the unit circle from $z = 1$ counter-clockwise by letting $z = e^{i\theta}$. What difficulty is encountered in trying to evaluate this integral by direct integration?

9.5.1. Determine an upper bound to the modulus of each of the following integrals:

(a) $4x+3iy$ integrated along the square $1, 1+i, i, 0$.

(b) $1/(1+z^2)$ integrated round the circle $|z| = R (> 1)$.

(c) $\sin z$ round each of the contours in (a) and (b).

(d) $z/(az^2+bz+c)$ (a, b, c complex) round a circle of large radius R. $(|a|R^2 > |b|R+|c|)$.

The bounds need not be the least possible. Comment on any difficulties.

Examine also the least upper bound of $|e^{iz}|$ (i) on the upper semicircle of radius R, (ii) on the lower semicircle.

CHAPTER 10

CAUCHY'S THEOREM. DERIVATIVES OF REGULAR FUNCTIONS

10.1. Integration along a Closed Contour

In the previous chapter we met several examples of integration along a path in which the result depended upon the end points only and not upon the path itself. Suppose that in Fig. 10.1

$$\int\limits_{ALB} f(z)\, dz = \int\limits_{AMB} f(z)\, dz.$$

Since it is easily shown (Exercise 10.1.1) that

$$\int\limits_{AMB} f(z)\, dz = -\int\limits_{BMA} f(z)\, dz,$$

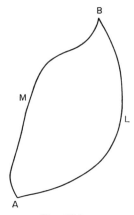

Fig. 10.1.

i.e. as in real variable, that reversing the direction of integration changes the sign of the integral, we may write instead

$$\int_{ALB} f(z)\,dz + \int_{BMA} f(z)\,dz = 0 \quad \text{or} \quad \int_{ALBMA} f(z)\,dz = 0,$$

where the integration has now been carried out round a closed contour C (i.e. a contour beginning and ending at the same point) beginning at A.

It is also easily shown that the value of this integral is not affected by the choice of the point on C at which integration begins (and therefore ends) (Exercise 10.1.2). Hence we may more simply write the result in the form

$$\int_C f(z)\,dz = 0.$$

A theorem known as Cauchy's theorem, which is of fundamental importance in complex variable theory, states that if f is *regular* in a domain D and if C lies entirely within D, then

$$\int_C f(z)\,dz = 0.$$

We shall re-word the theorem more specifically as follows.

THEOREM. (Cauchy's theorem.) *Let C be a closed contour which is simple (i.e. has no points of self-intersection) and let f be a function which is regular on C and everywhere inside C. Then*

$$\int_C f(z)\,dz = 0.$$

We note at this point that the functions we considered in § 9.3 whose integrals along paths from a to b depended only on a and b (and not on the choice of path) were regular in the whole plane [viz. $f(z) = z$, $f(z) = c$]. (Compare also Exercises 9.3.1 and 9.3.2.)

The full proof of Cauchy's theorem is somewhat tedious. We shall therefore limit ourselves at this point to providing a proof for the special case in which C is a triangle, and indicating how the result may be extended to more general contours. We shall also give a short proof, based on Stokes's theorem, applicable to general contours; unfortunately this proof is less satisfactory for our purposes in that it requires an additional assumption—which can, in fact, be deduced from Cauchy's theorem!

10.2. Cauchy's Theorem for a Triangle

Let f be regular at all points within and on a triangle ABC (Fig. 10.2). Let the length of the perimeter of ABC be L, and denote by Γ_0 the contour described by going once round the triangle in a counter-

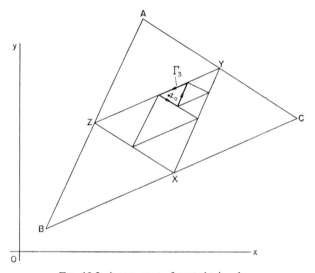

FIG. 10.2. A sequence of nested triangles.

clockwise direction. If $\int_{\Gamma_0} f(z)dz = 0$ there is nothing more to prove. We therefore suppose

$$\left| \int_{\Gamma_0} f(z)\ dz \right| = h > 0,$$

and we will deduce a contradiction.

We join the mid-points X, Y, Z of BC, CA, AB, respectively, to form four triangles, each of perimeter $L/2$, and observe that the sum of the integrals round these triangles, taken counter-clockwise, is equal to $\int_{\Gamma_0} f(z)\,dz$. This follows since all interior sides are included twice, having been traversed once in each direction, in the calculation of the sum of the integrals round the separate triangles.

From this we see that at least one of the four smaller triangles must give an integral of modulus greater than or equal to $h/4$, since the sum of the moduli of the integrals must be greater than or equal to the modulus of their sum, which is h. We select such a triangle, whose perimeter (counter-clockwise) we shall call the contour Γ_1. Joining the mid-points of the sides of Γ_1, we find, as before, a triangle Γ_2 of perimeter $L/2^2$ for which the modulus of the integral of $f(z)$ round the perimeter will be greater than or equal to $h/4^2$. This process can be repeated indefinitely, and after n repetitions we shall have selected a triangle Γ_n of perimeter $L/2^n$ for which

$$\left| \int_{\Gamma_n} f(z)\,dz \right| \geqslant h/4^n.$$

Now each triangle together with its interior is a closed and bounded set of points in the complex plane. We have constructed a nested sequence of these sets, and it is easy to show (see Exercise 10.2.1) that for such a nested sequence there must exist a point z_0 lying in all the members, i.e. in all the Γ_n. In particular, z_0 lies in $\Gamma_0 = ABC$, so f is regular at z_0. We may hence write

$$[f(z) - f(z_0)]/(z - z_0) = f'(z_0) + \phi(z),$$

where $|\phi(z)|$ approaches zero as z approaches z_0. [This is merely the definition of $f'(z_0)$.]

This expression may be rearranged to give an explicit expression for $f(z)$ in the form

$$f(z) = (z - z_0) f'(z_0) + \phi(z)(z - z_0).$$

Given an arbitrary $\varepsilon > 0$ we may select δ such that, for $|z - z_0| < \delta$, $|\phi(z)| < \varepsilon$. It is easily seen that for all points on the perimeter of

Γ_n, $|z-z_0| < L/2^n$; and we now select an N such that, for $n \geqslant N$, we have $L/2^n \prec \delta$. We note then that $|\phi(z)| < \varepsilon$ for all points z on the perimeters of the triangles concerned.

Let us consider the integral of $f(z)$ round Γ_N. We have

$$\int_{\Gamma_N} f(z)\, dz = f'(z_0) \int_{\Gamma_N} z\, dz - \int_{\Gamma_N} z_0\, f'(z_0)\, dz + \int_{\Gamma_N} \phi(z)\,(z-z_0)\, dz.$$

Now the first two integrals each vanish on a closed contour, as we saw in § 9.3. We hence have

$$\int_{\Gamma_N} f(z)\, dz = \int_{\Gamma_N} \phi(z)\,(z-z_0)\, dz,$$

whence, by (§ 9.5),

$$\left| \int_{\Gamma_N} f(z)\, dz \right| \leqslant \varepsilon\, L/2^N\, L/2^N = \varepsilon L^2/4^N.$$

Thus we have the inequalities

$$0 < h/4^N \leqslant \left| \int_{\Gamma_N} f(z)\, dz \right| \leqslant \varepsilon L^2/4^N,$$

from which we easily obtain

$$0 < h \leqslant \varepsilon L^2.$$

But ε is arbitrarily small, so by originally choosing $\varepsilon < h/L^2$ we have a contradiction, and we have therefore established that our hypothesis $h > 0$ is false. It follows that $h = 0$, proving the theorem for the case of a triangular contour.

The extension of the proof to any simple closed contour consisting of line segments (i.e. a polygonal contour) is quite obvious: it is merely necessary to divide the polygon into triangles and consider the sum of the integrals round the triangles, each of which would be zero. The completion of the proof of the general theorem consists in showing that in treating a "curved" contour as the limit of polygonal contours no residual error results, so that the theorem may be extended to general contours in a domain in which f is regular. (See Exercise 10.2.4.)

10.3. Proof of Cauchy's Theorem using Stokes's Theorem

Stokes's theorem, in its two-dimensional form, can be stated as follows.

THEOREM. *If $P(x, y)$, $Q(x, y)$, $\partial Q/\partial x$, $\partial P/\partial y$ are all continuous functions of x and y within a domain D bounded by a contour C, and also on C, then*

$$\int_C (P\,dx + Q\,dy) = \iint_D (\partial Q/\partial x - \partial P/\partial y)\,dx\,dy.$$

Now if $f(z) = u(x, y) + iv(x, y)$, we have by definition (see § 9.2)

$$\int_C f(z)\,dz = \int_C (u\,dx - v\,dy) + i \int_C (v\,dx + u\,dy).$$

Applying Stokes's theorem to each of these integrals in turn we easily obtain

$$\int_C f(z)\,dz = \iint_D (-\partial v/\partial x - \partial u/\partial y)\,dx\,dy + i \iint_D (\partial u/\partial x - \partial v/\partial y)\,dx\,dy,$$

where D is the interior of C.

Now by the Cauchy–Riemann equations, each of these integrands reduces identically to zero, and hence

$$\int_C f(z)\,dz = 0.$$

This brief and elegant proof unfortunately requires, for the validity of the double integration, the assumption of the continuity in D of $\partial u/\partial x$, etc.; or, what amounts to the same thing, that df/dz be continuous in D. We note that if f is differentiable, then

$$df/dz = \partial u/\partial x + i\,\partial v/\partial x = -i\,\partial u/\partial y + \partial v/\partial y.$$

However, we shall later use Cauchy's theorem itself to prove that f' is not only continuous, but even differentiable. It is this which led to our earlier remark that this simplified form of proof of the theorem is not completely satisfactory.

10.4. Cauchy's Integral

THEOREM. *If f is regular within and on a simple closed contour C, and if Z is any point within C, then*

$$f(Z) = \frac{1}{2\pi i} \int_C \frac{f(z)}{(z-Z)} \, dz.$$

In view of the regularity of the function f, the first point to note is why the integral does not reduce to zero by Cauchy's theorem. The reason is, of course, that the expression to be integrated is not even defined at $z = Z$, although regular elsewhere; and since $f(z)/(z-Z)$ approaches infinity as z approaches Z [because $f(z) \to f(Z)$ and $(z-Z) \to 0$], it is clear that there is in general no value which we could assign to $f(z)/(z-Z)$ at $z = Z$ to make it into a function regular or even continuous within and on C.

Proof. In order to prove the theorem, we construct another contour for which $z = Z$ will be an external point, so that Cauchy's theorem will be applicable to this new contour. The procedure is as follows. We first describe a small circle γ, lying entirely in the interior of C, with centre Z and radius r. From this circle we make a "cross-cut" to C. That is, we join the circle γ to C by a path with one end point on γ and the other on C, meeting γ and C nowhere else. We now consider the contour C' described as follows. Starting from any point on C, the curve C is traced counter-clockwise until the cross-cut is reached; the cut is followed in towards γ, which is then traversed in a clockwise direction until the end of the cut is again reached; the contour C' is completed by following the cut outwards until arriving at C and then following C again counter-clockwise to return to the starting-point (Fig. 10.3).

We would now like to apply Cauchy's theorem to C', regarding Z as an "exterior" point, but we cannot do this immediately since C' is not a simple contour (the whole cross-cut consisting of self-intersections). However, we can construct a sequence of simple closed

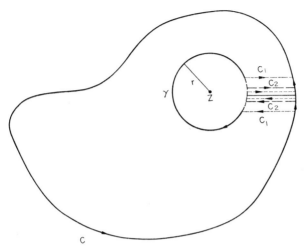

FIG. 10.3. Connecting contours by means of a cross-cut.

contours $C_1, C_2, C_3, \ldots$, approximating to C', by using paths joining γ to C, and lying very close to the cross-cut on either side rather than the cross-cut itself (Fig. 10.3). From the definition of integration it can be shown that for any function g regular on and near C',

$$\int_{C_n} g(z)\, dz \to \int_{C'} g(z)\, dz \quad \text{as} \quad n \to \infty.$$

In our case

$$\int_{C_n} \frac{f(z)}{z - Z}\, dz = 0$$

for each n by Cauchy's theorem since Z is an exterior point of C_n. Therefore

$$\int_{C'} \frac{f(z)}{z - Z}\, dz = 0.$$

In view of the fact that γ is traced in the negative direction, and that the two integrations along the cross-cut cancel each other, we at once obtain from this the result

$$\int_C \frac{f(z)}{(z-Z)}\, dz - \int_\gamma \frac{f(z)}{(z-Z)}\, dz = 0$$

or

$$\int_C \frac{f(z)}{z-Z}\, dz = \int_\gamma \frac{f(z)}{z-Z}\, dz.$$

We have thus reduced the problem to the consideration of the integral round the circle γ defined by $|z-Z| = r$. We may write this integral in the form

$$\int_\gamma \frac{f(z)}{z-Z}\, dz = \int_\gamma \frac{f(Z)}{z-Z}\, dz + \int_\gamma \frac{[f(z)-f(Z)]}{z-Z}\, dz$$

$$+ I_1 + I_2, \quad \text{say.}$$

We will evaluate the first integral and show that the second is zero. (This device, as will be found, is quite common in contour integration). Before proceeding we need to note a simple result as follows. If

$$w = e^{i\theta} = \cos\theta + i\sin\theta,$$

then

$$dw/d\theta = -\sin\theta + i\cos\theta$$
$$= i(i\sin\theta + \cos\theta)$$
$$= i(\cos\theta + i\sin\theta)$$
$$= ie^{i\theta}.$$

We can now evaluate I_1 by means of the substitution $z = Z + re^{i\theta}$, using θ as the parameter. Since $f(Z)$ is merely a constant, we obtain

$$I_1 = f(Z) \int_0^{2\pi} \frac{rie^{i\theta}}{re^{i\theta}}\, d\theta = f(Z)\int_0^{2\pi} i\, d\theta = 2\pi i f(Z).$$

We now consider I_2. We can at once assert that

$$|I_2| = \left| \int_{\gamma} \frac{f(z) - f(Z)}{z - Z} \, dz \right|$$

$$\leqslant |f(Z) - f(z)|_{\substack{\max \\ z \text{ on } \gamma}} (1/r) 2\pi r \quad \text{(by § 9.5)}$$

(since $|1/(z-Z)| = 1/r$ on γ, and the length of the circumference of γ is $2\pi r$),

$$= 2\pi |f(z) - f(Z)|_{\substack{\max \\ z \text{ on } \gamma}}$$

Now by the continuity of f, it follows that if we take a circle γ of sufficiently small radius r, the difference between $f(z)$ and $f(Z)$ may be made as small as we please, and hence I_2 may be made arbitrarily small. More precisely, given any number $\varepsilon > 0$ we can choose r to make $|I_2| < \varepsilon$. But the integral round C is equal to $I_1 + I_2$ and is independent of r. We have calculated I_1 and seen that it is independent of r. It follows that I_2 also must be independent of r. Since its modulus is less than any given positive ε, it must be zero exactly. We are therefore left with the result

$$\int_{C} \frac{f(z)}{z - Z} \, dz = 2\pi i f(Z),$$

or, rearranging,

$$f(Z) = \frac{1}{2\pi i} \int_{C} \frac{f(z)}{z - Z} \, dz.$$

This proves the theorem.

10.5. The Derivative of a Regular Function

Cauchy's integral enables us to express the derivative of a regular function at any point in the form of an integral round a suitable contour; a form that will prove exceedingly useful in the next chapter.

We have, by definition,

$$f'(Z) = \lim_{h \to 0} \frac{f(Z+h)-f(Z)}{h}$$

$$= \lim_{h \to 0} \frac{1}{h}\left[\frac{1}{2\pi i}\int_C \frac{f(z)}{z-Z-h}\,dz - \frac{1}{2\pi i}\int_C \frac{f(z)}{z-Z}\,dz\right],$$

where C is a simple closed contour with f regular on and inside C, the point Z lies in C, and $Z+h$ also lies in C for all the values of h considered.

We write this as

$$f'(Z) = \lim_{h \to 0} \frac{1}{2\pi i h}\int \frac{hf(z)}{(z-Z)(z-Z-h)}\,dz$$

$$= \lim_{h \to 0} \frac{1}{2\pi i}\int_C \frac{f(z)}{(z-Z)(z-Z-h)}\,dz.$$

It appears probable that this limit is

$$\frac{1}{2\pi i}\int_C \frac{f(z)}{(z-Z)^2}\,dz$$

although this cannot be assumed, since we require the limit of the integral, which is not necessarily the same as the integral of the limit. (See Exercise 10.5.1.) We therefore write

$$f'(Z) = \lim_{h \to 0} \frac{1}{2\pi i}\left[\int_C \frac{f(z)}{(z-Z)^2}\,dz \right.$$

$$\left. + \int_C f(z)\left(\frac{1}{(z-Z)(z-Z-h)} - \frac{1}{(z-Z)^2}\,dz\right)\right]$$

The first integral is clearly independent of h. We shall examine the second and hope to show that it approaches zero as $h \to 0$. The expression in the brackets reduces to $h/(z-Z)^2(z-Z-h)$, so that if we denote the integral by I, we have by § 9.5,

$$|I| \leqslant |f(z)|\,\max_{z \text{ on } C}\,(|h|/|z-Z|^2\,|z-Z-h|)\,\max_{z \text{ on } C}\,\times \text{length of } C.$$

Since Z is an internal point of the contour, the lower bound of the distances of points on C from Z (which will actually be attained for some point or points on C) will be positive (i.e. not zero). Let us call this distance d. Since h is to approach zero, we may also assume that $|h| < \frac{1}{2}d$ (Fig. 10.4). Then, writing M for $|f(z)|_{\substack{\max \\ z \text{ on } C}}$, which will

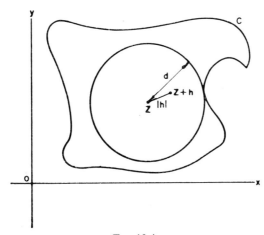

FIG. 10.4.

certainly be finite in view of the regularity, and hence continuity, of f in and on C, we obtain

$$|I| < \frac{M|h|}{d^2 \frac{1}{2} d} L,$$

where L is the length of C.

Evidently this approaches zero with h, and so we have proved the required result, namely

$$f'(Z) = \frac{1}{2\pi i} \int_C \frac{f(z)}{(z-Z)^2} dz.$$

We note that this is precisely the result obtained by differentiating the integral expression for $f(Z)$ with respect to Z under the integral sign. Applying this procedure does not, of course, provide a proof of the result, since such a procedure itself would have to be justified.

10.6. Repeated Differentiation

With the same notation as in the previous section, and by means of a similar procedure, we easily establish that the second derivative of $f(z)$ with respect to z may be written in the form

$$f''(Z) = \frac{2}{2\pi i} \int_C \frac{f(z)}{(z-Z)^3}\,dz + R,$$

where $R = \lim_{h \to 0} \frac{1}{2\pi i} \int_C f(z) \left[\frac{2z-2Z-h}{(z-Z)^2(z-Z-h)^2} - \frac{2}{(z-Z)^3} \right] dz.$

On bringing to a common denominator, the terms in square brackets reduce to

$$h[3(z-Z)-2h]/(z-Z-h)^2(z-Z)^3,$$

whence the modulus of the entire expression is less than

$$(\tfrac{1}{2}\pi)\, M\, |h|\, (3d+d)/(\tfrac{1}{2}\,d)^2 d^3,$$

which evidently vanishes as h approaches zero, so that $R = 0$. We have hence shown that f' is regular wherever f is regular, and that its value at a point Z is that which would have been obtained by direct differentiation of Cauchy's integral under the integral sign.

This result suggests that, at all points Z within the domain where f is regular, the function f has derivatives of all orders given by the formula

$$f^{(n)}(Z) = \frac{n!}{2\pi i} \int_C \frac{f(z)}{(z-Z)^{n+1}}\,dz.$$

This result, proved straightforwardly by mathematical induction, is left as an exercise (10.6.1).

Note that it follows that all the derivatives of a regular function are themselves regular at all points within the domain in which the original function was regular.

We continue this chapter with a theorem which is, in a sense, a converse of Cauchy's theorem.

10.7. Morera's Theorem

A converse to Cauchy's theorem would say that if f is defined in a domain D and if $\int_C f(z)\, dz$ exists and is zero for every simple closed contour in D, then f must be regular. This, however, is easily seen to be false; take for instance D to be the whole plane and f to be zero everywhere except at 0, but having the value 1 at 0. Then $\int_C f(z)\, dz = 0$ for all C; but f cannot be regular since it is not even continuous. However, if we assume originally that f is continuous then the theorem does become true.

THEOREM (Morera's theorem). *If f is continuous in a domain D, and if $\int_C f(z)\, dz = 0$ for every closed contour C in D, then f is regular in D.*

In order to prove the theorem we need a lemma. We have already seen (§ 10.1) that the vanishing of the integral round all closed contours is equivalent to the fact that integration between any two points is independent of the path of integration. Hence, if we take an arbitrary point a in D as the initial point of a path of integration we can define a function F by

$$F(z) = \int_a^z f(Z)\, dZ,$$

without specifying the path. The lemma states:

LEMMA. *$F(z)$ as defined above is differentiable, and $F'(z) = f(z)$.*

Proof of lemma. Let us consider the following paths of integration; C_1, from a to z, C_2 the straight line from z to a neighbouring point $z+h$, and C_3, from a to $z+h$ not passing through z; where a, z, $z+h$, *and the paths* are all in D (Fig. 10.5). Then, by our assumption,

$$\int_{C_1} f(Z)\, dZ + \int_{C_2} f(Z)\, dZ = \int_{C_3} f(Z)\, dZ.$$

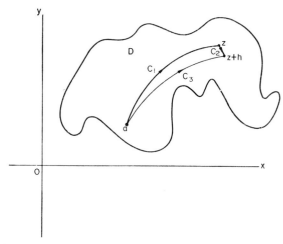

FIG. 10.5. Morera's theorem. Integration along paths lying within D.

In attempting to find the derivative of $F(z)$, if it exists, we have to examine the limit as h approaches zero of the expression

$$\frac{1}{h}[F(z+h)-F(z)] = \frac{1}{h}\left(\int_{C_3} f(Z)\,dZ - \int_{C_1} f(Z)\,dZ\right)$$

$$= \frac{1}{h}\int_{C_2} f(Z)\,dZ = \frac{1}{h}\int_{C_2} f(z)\,dZ + \frac{1}{h}\int_{C_2} [f(Z)-f(z)]\,dZ$$

$$= \frac{1}{h}f(z)\int_{C_2} 1\,dZ + I, \quad \text{say,}$$

$$= (1/h)f(z)\left[Z\right]_z^{z+h} + I$$

$$= (1/h)f(z)\,h + I,$$

$$= f(z)+I.$$

Now

$$|I| \leqslant (1/|h|)|f(Z)-f(z)|_{\substack{\max \\ z \text{ on } C}} \times |h| \quad \text{by (§ 9.5)}$$

$$= |f(Z)-f(z)|_{\substack{\max \\ z \text{ on } C}}$$

which approaches zero as h tends to zero by the continuity of f at Z. Hence

$$\lim_{h \to 0} [F(z+h)-F(z)]/h = f(z) \quad \text{or} \quad F'(z) = f(z).$$

The proof of Morera's theorem now follows at once.

Proof. We have proved that f is the derivative of F, which means that F is regular. But f is the derivative of a regular function, and, therefore (see previous section), f itself must be regular.

10.8. Liouville's Theorem

It is comparatively easy to find functions of a real variable x which are both bounded and differentiable for all values of x from $-\infty$ to $+\infty$. For example, $f(x) = 1/(x^2+a^2)$, $(a \neq 0)$, $f(x) = \sin x$. As it happens, neither of these functions is bounded if complex values of the variable are permitted—the modulus of the first expression increases without limit as z approaches ia (or $-ia$), while if for x we substitute iy, the modulus of the second approaches infinity with y since $\sin iy = i \sinh y = \frac{1}{2}i(e^y - e^{-y})$, etc.

While it is simple enough to devise functions of a complex variable that are bounded over the whole complex plane, e.g. $f(z) = f(x+iy) = 0$, when x and/or y is rational; $f(z) = 1$, otherwise, any attempt to find a non-trivial function of z which is both regular and bounded over the entire plane invariably fails. The explanation of this is contained in the following theorem.

THEOREM (Liouville's theorem). *If f is regular and bounded throughout the complex plane, then $f(z)$ is a constant.*

Proof. (See also Exercise 10.8.1.) Let $|f(z)| \leqslant M$ for all z, and take any two points z_1, z_2. Let C be a circle with centre the origin and radius R greater than $|z_1|$ and $|z_2|$. Consider the difference in the values of the function at z_1 and z_2. By Cauchy's integral formula, we have

$$f(z_1) - f(z_2) = \frac{1}{2\pi i} \int_C \frac{f(z)}{z - z_1} \, dz - \frac{1}{2\pi i} \int_C \frac{f(z)}{z - z_2} \, dz$$

$$= \frac{1}{2\pi i} \int_C f(z) \left(\frac{1}{z - z_1} - \frac{1}{z - z_2} \right) dz$$

$$= \frac{(z_1 - z_2)}{2\pi i} \int_C \frac{f(z)}{(z - z_1)(z - z_2)} \, dz,$$

whence

$$|f(z_1) - f(z_2)| \leqslant \frac{|z_1 - z_2|}{2\pi} \frac{M}{(R - |z_1|)(R - |z_2|)} \, 2\pi R$$

$$= \frac{M |z_1 - z_2|}{R(1 - |z_1|/R)(1 - |z_2|/R)},$$

which approaches zero as R approaches infinity. However, $|f(z_1) - f(z_2)|$ is clearly independent of R. Thus $f(z_1) = f(z_2)$, where z_1 and z_2 are any two points. It follows that $f(z)$ must be constant, which was what we wished to prove.

Exercises

10.1.1. Prove that if the path of integration is reversed the sign of the integral is changed.

10.1.2. Prove that the value of an integral evaluated round a closed contour is unaffected by changing the point at which integration begins and ends.

10.2.1. On what theorem do we base the assertion that a nested sequence of triangles will contain a point common to all of them? (See § 2.4, and remember that a closed set contains its limit point.) Prove the result formally.

10.2.2. Evaluate $\phi(z)$ (defined as in § 10.2) explicitly in terms of z and z_0 for the cases $f(z) =$ (a) z^2; (b) z^3; (c) $1/z$ ($z_0 \neq 0$).

10.2.3. Prove that the distance from any interior point in a triangle to any point on the perimeter is less than the length of the perimeter. Show that it is also (a) less than the sum of any two sides, and (b) less than half the perimeter.

10.2.4. Let f be regular on and inside a simple closed contour C. Let Q be a fixed large square containing C. We know the following facts: (i) if f is regular at z_0 then it is regular in a small neighbourhood of z_0 (from the definition of regular); (ii) if f is regular at z_0 then for all z sufficiently close to z_0 we have

$$f(z) - f(z_0) = f'(z_0)(z - z_0) + \varepsilon \, | z - z_0 |$$

where $\varepsilon \to 0$ as $z \to z_0$ [from the definition of $f'(z_0)$].

Assume now the following facts: (iii) the square Q can be divided into small squares by lines parallel to the sides such that f is regular in every square which meets C or lies in the interior of C, of which there are only a finite number. Call these squares $S_1, \ldots, S_n$; (iv) given any $\varepsilon > 0$ the squares in (iii) can be chosen such that in each S_r there exists a point z_r with

$$f(z) - f(z_r) = f'(z_r)(z - z_r) + \varepsilon_r \, | z - z_r |,$$

where z lies in S_r, and such that each ε_r is less than the given ε.

Prove:

(a) If S_r has side a, and the perimeter of S_r taken in a counter-clockwise direction is denoted by C_r, then

$$\left| \int_{C_r} | z - z_r | \, dz \right| < 4 \sqrt{2} \, a^2.$$

(b) $$\int_Q f(z) \, dz = \sum_{r=1}^{n} \int_{C_r'} \varepsilon_r | z - z_r | \, dz,$$

where C_r' denotes C_r if S_r lies entirely inside C, and C_r' denotes one or more contours made up of pieces of C_r (call them C_{r1}) and pieces of C (call them C_{r2}) lying in S_r if S_r meets C.

(c) $$\sum \left| \int_{C_{r1}} \varepsilon \, | z - z_r | \, dz \right| < \varepsilon 4 \sqrt{2} \, A,$$

where A is the area of the square Q.

(d) $$\sum \left| \int_{C_{r2}} \varepsilon \, | z - z_r | \, dz \right| < \varepsilon L \sqrt{2} \, S,$$

where L is the length of the side of Q and S is the length of C.

Deduce Cauchy's theorem for the contour C from the above results.

10.3.1. Verify the equivalence of the path and area integrals as given by Stokes's theorem, for the following cases:

(a) $P = y$, $Q = x$, C being the circle $x^2 + y^2 = a^2$;

(b) $P = y$, $Q = 0$, C as in (a).

10.3.2. Prove that if $f(z) = u(x, y) + iv(x, y)$ is twice differentiable, then all the second-order derivatives $\partial^2 u / \partial x^2$, etc., exist. Prove further that if $f^{(n)}$ exists in D for all n, then partial derivatives of u and v of all orders exist in D and are continuous.

Show by means of an example that it is possible for a real function to have a derivative at a given point, while its derivative is not itself differentiable at the point.

10.4.1. Evaluate by means of Cauchy's integral formula:

(i) $\int_C dz/z$, where C is a simple closed contour containing the origin;

(ii) $\int_C dz/(z-2)$, where C is the circle $|z| = 3$, taken in a counter-clockwise sense. Evaluate (ii) also for the case in which C is the unit circle.

(iii) $\int_C (2z-3)\,dz/(z+1)(z-4)$, where C is the circle $|z| = 2$.

Evaluate the integral also (a) where C is $|z-2| = 1$, (b) where C is $|z-4| = 2$. Unless otherwise stated, all contour integrals are considered to be taken in a counter-clockwise direction. (Hint: $(2z-3)/(z-4)$ is a regular function of z except at $z = 4$; $(2z-3)/(z+1)$ is regular except at $z = -1$.)

10.5.1. By considering the integral between 0 and 1 of $2nxe^{-nx^2}$, and using the fact that $\lim_{n \to \infty} n/e^n$ approaches zero, show that the limit of an integral is not necessarily equal to the integral of the limit. Find other examples.

10.5.2. By means of the formula for the derivative of f in the form of an integral, evaluate

$$\int \sin z\,dz/(z+i)^2,$$

where C is the circle $|z| = 2$. Construct and evaluate some other simple contour integrals of this sort. (Hint: f needs to be a function that can be differentiated, and the derivative evaluated at a given point within the contour.)

10.6.1. Prove the formula of § 10.6 for the nth derivative at Z of the regular function f.

10.6.2. Use the procedure of Exercise 10.5.2 in order to evaluate

$$\int_C \frac{e^{2z}}{(z-\pi i)^3}\,dx,$$

where C is any contour with πi as an interior point. Examine further possibilities considering derivatives of order higher than 2.

10.6.3. Establish Cauchy's inequality for $f^{(n)}(Z)$, i.e.

$$|f^{(n)}(Z)| \leqslant Mn!/r^n,$$

where f is regular within and on the circle $|z-Z| = r$, and n is any non-negative integer, whilst M denotes the maximum value of $|f(z)|$ on the circle.

10.7.1. Evaluate $\int_C \frac{dz}{Z^2}$, where C is (i) a simple closed contour not containing the origin; (ii) a circle with centre the origin; (iii) any simple closed contour containing the origin in its interior. (Use an argument as in § 10.4.) Comment on the results of (i) and (iii) in relation to Morera's theorem.

10.8.1. By using Cauchy's inequality (Exercise 10.6.3) for the case $n = 1$, and letting the radius r approach infinity, deduce that if f is a function regular over the whole plane, then $f'(Z) = 0$ for all points Z in the z-plane, i.e. that $f'(z) = 0$. Hence prove Liouville's theorem.

10.8.2. Point out why none of the following functions are bounded over the complex plane. State any points at which they are not defined:

(a) $f(z) = e^{1/(z+a^2)}$ (a real);

(b) $1/\sqrt{(z^2+b^2)}$ (b real);

(c) $\cos(1/z)$;

(d) $\tanh z$.

10.8.3. Suppose that $f(z)$ is a polynomial $a_0 + \ldots + a_n z^n$, ($n \geqslant 1$, $a_n \neq 0$), and that $|f(z)|_{\min} = k$. By considering the modulus of the function $F(z) = 1/f(z)$ and using Liouville's theorem, deduce that $k = 0$. This proves the "fundamental theorem of algebra" that every polynomial other than a constant has a zero in the complex plane. (See § 14.4.)

CHAPTER 11

TAYLOR'S THEOREM
AND LAURENT'S THEOREM

11.1. A Useful Identity

The two very important theorems which form the subject of this chapter depend for their proofs on Cauchy's integral formula and on an algebraic result which we will prove first.

The geometric progression $1+c+c^2+\ldots+c^{n-1}$, where c is any complex number, has the sum $(1-c^n)/(1-c)$, which may be written in the form $1/(1-c)-c^n/(1-c)$. We may rearrange this result in the form

$$1/(1-c) = 1+c+c^2+\ldots+c^{n-1}+c^n/(1-c),$$

which could also have been obtained by direct division.

Using this identity we may write

$$1/(z-Z) = 1/[(z-a)-(Z-a)] = [1/(z-a)]\,[1/\{1-(Z-a)/(z-a)\}]$$

$$= [1/(z-a)]\left(1+\frac{Z-a}{z-a}+\left(\frac{Z-a}{z-a}\right)^2+\ldots+\left(\frac{Z-a}{z-a}\right)^{n-1}\right.$$

$$\left.+\left(\frac{Z-a}{z-a}\right)^n\left(\frac{1}{1-\left(\frac{Z-a}{z-a}\right)}\right)\right),$$

which readily simplifies to

$$1/(z-Z) = 1/(z-a)+(Z-a)/(z-a)^2+\ldots+(z-a)^{n-1}/(Z-a)^n$$

$$+\left(\frac{Z-a}{z-a}\right)^n\left(\frac{1}{z-Z}\right).$$

This is the result which we shall need later.

11.2. Taylor's Theorem

THEOREM. *Let f be regular within the circle* $|z-a| = R$ *and let Z be any point in the interior of the circle. Then*

$$f(Z) = \sum_{n=0}^{\infty} a_n(Z-a)^n,$$

where $a_n = f^{(n)}(a)/n!$.

Proof. In order to prove the theorem, we consider contour integration round a circle C with centre a and radius R_1, where $|Z-a|$ $= r < R_1 < R$. Evidently f is regular in and on this contour, and we may apply Cauchy's theorem and his integral formula.

We take the result of the preceding section (§ 11.1), multiply each term by $f(z)/2\pi i$, and integrate round C with respect to z, taking Z and a in the formula of § 11.1 to be the points Z, a in the statement of the theorem.

This gives

$$(1/2\pi i) \int_C \frac{f(z)}{z-Z} dz = (1/2\pi i) \int_C \frac{f(z)}{z-a} dz + [(Z-a)/2\pi i] \int_C \frac{f(z)}{(z-a)^2} dz$$

$$+ \ldots + [(Z-a)^{n-1}/2\pi i] \int_C \frac{f(z)}{(z-a)^n} dz + I_n,$$

where

$$I_n = \frac{(Z-a)^n}{2\pi i} \int_C \frac{f(z)}{(z-a)^n(z-Z)} dz.$$

Leaving consideration of I_n for the moment, we see that the other integrals may all be evaluated by means of the results of §§ 10.4, 10.5, and 10.6, leading to

$$f(Z) = f(a) + (Z-a)f'(a) + \ldots + (Z-a)^{n-1}f^{(n)}(a)/(n-1)! + I_n.$$

In order to complete the proof of the theorem, all we need to show is that I_n approaches zero as n tends to infinity.

Using § 9.5 we have

$$|I_n| \leqslant \frac{r^n}{2\pi} \frac{M}{R_1^n(R_1-r)} 2\pi R_1,$$

where M is the maximum value of $|f(z)|$ on the circle.

This value is necessarily finite, since $f(z)$ is defined, and in fact regular (and therefore continuous) at all points of C. Thus

$$|I_n| \leqslant A(r/R_1)^n,$$

where $A = R_1 M/(R_1-r)$, which is independent of n.

Since $r < R_1$ it now follows at once that I_n approaches zero as n approaches infinity; this completes the proof of the theorem.

This theorem shows that if f is regular in the interior of any circle C, its value at any interior point of C may always be expressed as a convergent power series. Now if a function f is regular at a point z_0 it is also (from the definition) regular in some neighbourhood of z_0, and hence in the interior of some small circle C with centre at z_0. For all z inside C, $f(z)$ can, as we have seen, be expressed as a convergent power series. Earlier (§ 5.11) we proved conversely that any power series sums to give a regular function within its radius of convergence. Representing regular functions by power series and thinking of power series as regular functions frequently proves the best way of dealing with these functions.

11.3. Radius of Convergence of the Taylor Series

Suppose we wish to express $f(z)$ as a Taylor series in powers of $z-z_0$, in the neighbourhood of the point z_0. It will be important to know the radius of the circle of convergence of this series, i.e. how far away from z_0 the power series for f will be valid.

Let $z = z_1$ be the nearest singularity to z_0, i.e. the nearest point to z_0 at which f is either not regular or not defined. Then, evidently, the function is regular in the interior of a circle $|z-z_0| = r$ provided that $r < |z_1-z_0|$, and it may thus be expressed there in series form. On the other hand, if $r > |z_1-z_0|$ the function is not regular throughout the interior of the circle, and thus cannot be expressed as a power series convergent at all points within the circle. We deduce from this that the radius of convergence we seek is precisely the distance from z_0 to the nearest singularity.

This consideration frequently provides us with the simplest method of determining the circle of convergence.

Notice that there may well be a point z_2 inside $|z-z_0| = r$ such that for some radius r' the circle $|z-z_2| = r'$ lies partly inside and partly

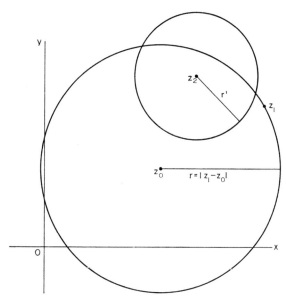

Fig. 11.1. Analytic continuation.

outside $|z-z_0| = r$ but, nevertheless, contains no singularities of f (Fig. 11.1).

In this case we can extend the region in which f is expressible as a power series by also using power series with the new point z_2 as origin. This process is known as *analytic continuation*.

11.4. Uniqueness of Power Series

Very probably in practice the method of continued differentiation in order to obtain successive coefficients $f(a), f'(a), f''(a), \ldots, f^{(n)}(a),$ $\ldots$, is not the easiest way to obtain a power series to represent a given function. For example, in the case $f(z) = 1/(1-z)$ already considered, both mechanical division and use of the binomial series are applicable as alternatives, giving directly the series $1+z+z^2+ \ldots$. The question then evidently arises whether a power series representing a given function, when obtained by whatever method, is unique. If this is the case, then "short cuts", when they are available, may be used with impunity—the series obtained will be identical to the Taylor series.

If the series is in powers of, say, $(z-a)$, we see by replacing $(z-a)$ by Z throughout that there will be no loss of generality if we limit ourselves to proving the following theorem.

THEOREM. *Suppose*

$$\sum_{r=0}^{\infty} a_r z^r \quad and \quad \sum_{r=0}^{\infty} b_r z^r$$

both converge for $|z| < \varrho$ *(for some* $\varrho > 0$*) and suppose that*

$$\sum_{r=0}^{\infty} a_r z^r = \sum_{r=0}^{\infty} b_r z^r$$

for all such z. Then $a_r = b_r$ *(r = 0, 1, 2, \ldots).*

Proof. We may, in view of the convergence of the series concerned, write what we have to prove in the equivalent form. If

$$\sum_{r=0}^{\infty} (a_r - b_r) z^r = 0$$

for all z such that $|z| < \varrho$, then $a_r = b_r$ for all r. By letting $z = 0$ we see at once that $a_0 = b_0$. Also, since a power series may be differentiated as many times as we please within its circle of convergence, we obtain after n such differentiations

$$0 = n! \, (a_n - b_n) + (n+1)! \, (a_{n+1} - b_{n+1})z/1!$$
$$+ (n+2)(a_{n+2} - b_{n+2})z^2/2! + \dots.$$

Letting $z = 0$ gives at once $a_n - b_n = 0$ or $a_n = b_n$ ($n = 1, 2, 3, \dots$). This was what we wished to prove.

11.5. Maclaurin's Series

If the origin is taken as the centre of the circle C of integration used in obtaining the Taylor series of a function f (see § 11.2), then we shall have $f(z)$ expressed as a power series in z. This special form of Taylor's series, where $a = 0$, is known as Maclaurin's series. We have already met series for exp z, sin z, log $(1+z)$, and some other simple functions, and by the uniqueness proved above these must be the Maclaurin series for these functions. The general expression for a Maclaurin series of a function f regular at the origin is evidently

$$f(z) = \sum_{n=0}^{\infty} f^{(n)}(0)z^n/n!$$

11.6. Laurent's Theorem

Suppose that f is not defined, or is not regular, at a point a. Then we shall be unable to express it in the neighbourhood of a as a convergent series of positive powers of $(z-a)$, for if we could do so then by § 5.11 it would be regular at a. However, the following theorem often enables us, in this case, to obtain a series expansion involving both positive and negative powers of $(z-a)$, valid near a (although of course not at a itself).

THEOREM (Laurent's theorem). *Let C_1 and C_2 be two concentric circles, with centre a and radii r_1 and r_2 respectively, where $r_1 < r_2$. (We admit the case $r_1 = 0$, thinking of the point a as a "degenerate"*

circle.) *Let f be regular on the circles and within the annulus between*
them. Then if Z is any point within the annulus, f(Z) can be expressed
in the form

$$f(z) = \sum_{n=0}^{\infty} a_n(Z-a)^n + \sum_{n=1}^{\infty} b_n(Z-a)^{-n},$$

where each of the infinite series is convergent and where

$$a_n = (1/2\pi i) \int_C \frac{f(z)}{(z-a)^{n+1}} \, dz \quad (n = 0, 1, 2, \ldots),$$

$$b_n = (1/2\pi i) \int_C (z-a)^{n-1} f(z) \, dz \quad (n = 1, 2, 3, \ldots).$$

The proof of the theorem depends on the two identities:

$$1/(z-Z) = 1/(z-a) + (Z-a)/(z-a)^2 + \ldots$$
$$+ (Z-a)^{n-1}/(z-a)^n + (Z-a)^n/(z-a)^n(z-Z) \quad \text{(A)}$$

and

$$-1/(z-Z) = 1/(Z-a) + (z-a)/(Z-a)^2 + \ldots$$
$$+ (z-a)^{n-1}/(Z-a)^n + (z-a)^n/(Z-a)^n(Z-z). \quad \text{(B)}$$

The former, (A), was proved in § 11.1, and (B) is derived immediately
from (A) by interchanging z and Z.

By means of a cross-cut K between C_1 and C_2, we easily obtain, by
considering the contour C', consisting of C_2 followed by K followed
by C_1 reversed followed by K reversed, and applying Cauchy's integral
formula (compare the details of § 10.4):

$$f(Z) = (1/2\pi i) \int_{C_2} \frac{f(z)}{z-Z} \, dz - (1/2\pi i) \int_{C_1} \frac{f(z)}{z-Z} \, dz.$$

The first integral, along C_2, may be converted by means of identity
(A) to the form of an infinite series, precisely as has already been done

in proving Taylor's theorem—with one important difference: in this case we are no longer able to replace the expression

$$(1/2\pi i) \int_{C_2} \frac{f(z)}{(z-a)^{n+1}} dz$$

by $n!f^{(n)}(a)$, since $f^{(n)}(a)$ may well not even exist, a not being a point within the region between the circles where f is given to be regular.

In order to deal with the second integral, we transform it by means of (B) above to obtain

$$-(1/2\pi i) \int_{C_1} \frac{f(z)}{z-Z} dz$$

$$= (1/2\pi i) \int_{C_1} \frac{f(z)}{z-a} dz + \ldots + [(Z-a)^{-n}/2\pi i] \int_{C_1} (z-a)^{n-1} f(z) dz + J_n,$$

where

$$J_n = (1/2\pi i) \int_{C_1} \frac{f(z)(z-a)^n}{(Z-a)^n(Z-z)} dz.$$

By referring to the definition of the b_n we see that the theorem is now proved if we can show that J_n tends to zero as n approaches infinity. In fact, writing $|Z-a| = r$ and letting M be the maximum value of $|f(z)|$ on C_1, we have

$$|J_n| \leqslant (1/2\pi) Mr_1^n/r^n(r-r_1) \cdot 2\pi r_1 = K(r_1/r)^n,$$

where K is independent of n. Since $r_1 < r$, the result follows.

Exercises

11.1.1. Find the sum to n terms of the series

$$1 + 2z + 3z^2 + 4z^3 + \ldots + rz^{r-1} + \ldots.$$

By replacing z by $(Z-a)/(z-a)$ express $1/(z-Z)^2$ in the form of a finite power series and a remainder term, using powers of $(Z-a)$.

190   COMPLEX VARIABLES

11.1.2. Express $1/(z-Z)$ in the form of a finite series of negative powers of $(Z-a)$, together with a remainder term.

11.2.1. In the proof of Taylor's theorem, state why a circle interior to the given circle $|z-a| = R$ is used as the contour for integration purposes, rather than the given circle itself. Under what condition could the original circle have been used. Why do you think the theorem was not stated in a form in which this condition was satisfied?

11.2.2. By means of Taylor's theorem, express each of the following functions of z in the form of an infinite series of positive powers of $(z-a)$: state in each case any limitations on the value of a, and also the radius of convergence of the series: (i) $1/z$; (ii) $1/(z-b)$; (iii) e^z; (iv) $\log z$; (v) $\sin kz$; (vi) z^n (n not an integer).

11.2.3. Obtain by means of Taylor's theorem the first three non-zero terms in the expansions, in powers of the given expression, of the following functions:

(i) $f(z) = \tan^{-1} z$ in powers of $(z-1/4)$;
(ii) $f(z) = \log (z-3)$ in powers of $(z-2i)$;
(iii) $f(z) = e^{\sin z}$ in powers of z.

11.3.1. If $1/(z^2+16)$ is expanded in powers of $(z-3)$ by Taylor's theorem, find the radius of convergence of the series.

11.3.2. Determine the radii of convergence of the power series expansions, derived from Taylor's theorem, of the following expressions:

(i) $1/(z-2)(z^2+1)$ in powers of (a) z; (b) $z+5$; (c) $z-1$;
(ii) $\sin 1/z$ in powers of $(z+1)$;
(iii) $\tan^{-1} [1/(1+z^2)]$ in powers of $(z-2i)$.

11.4.1. Obtain by any convenient method the following expansions, and state in each case the radius of convergence of the series:

(i) $\sinh z$ in powers of (a) z; (b) $(z-\pi i)$;
(ii) $1/(z-2)(z+3)$ in powers of (a) $(z+1)$; (b) $(z+2)$; (c) $(z-5)$.

11.4.2. From the series for $\cos z$ in powers of z, obtain the first three non-zero terms of the power series for $\sec z$ in powers of z in each of the following four ways.

(i) direct long division (of 1 by the first few terms of the $\cos z$ series);
(ii) assuming a solution of the form $a_0+a_1z+a_2z^2+ \ldots$, find the first few terms of the product series $\cos z \sec z$, and equate the constant term to 1 and all other coefficients to zero;
(iii) the use of Taylor's theorem and continued differentiation;
(iv) using the relationship $\sec z = 2 \exp iz/(\exp 2iz+1)$.

11.5.1. Find the Maclaurin series for each of the following: (a) $\sin z^2$; (b) $\sin 2z$; (c) $\log (1+iz)$; (d) $\sinh cz$; (e) $z^2/(1+z^2)$; (f) $z^2/(1+z)^2$; (g) $\tan^{-1} az$; (h) $\log (1+z^2)$. (Hint: use any method—do not differentiate repeatedly if this method becomes tedious.)

11.6.1. Find a suitable upper bound M_n for the modulus of the remainder term J_n in Laurent's series for the function $f(z) = \sin z/(z+\frac{1}{2}i)$, where the annulus is between the circles $|z| = 1$, $|z| = 2$. ("Suitable" in this context means that M_n approaches zero as n approaches infinity.)

11.6.2. Express the function $f(z) = z/(2z+1)(z-2)$ as a power series in (a) positive powers of z, and (b) negative powers of z, determining in each case the domain of convergence. Find also a suitable series for points in the interior of the annulus between these domains.

11.6.3. Find a Laurent series in positive and/or negative powers of z for cosec z. (Give the first three terms in ascending powers of z.)

11.6.4. Find Taylor and Laurent series for $f(z)$ in powers of z when $f(z)$ is (a) $(z-1)/(z-2)$, (b) $(z-1)/(z-2)(z+3)$, stating in each case the domains in which the series are valid. [Give two domains for (a), three for (b). (Hint: (b) may be written in the forms $(1-1/z)/z(1-2/z)(1+3/z)$, $(1-z)/2(1-z/2)3(1+z/3)$, $(z-1)/z(1-2/z)(3+z)$].

11.6.5. Obtain the first four non-zero terms of the Laurent series for (a) $\sin z/z^3$ in powers of z, and (b) $e^z/(1+z^2)$ in powers of $(z-i)$, in ascending order starting with the greatest negative power.

CHAPTER 12

ZEROS AND SINGULARITIES. MEROMORPHIC FUNCTIONS

12.1. Zeros

DEFINITION. *If a function f is regular in a domain containing the point z_0, and if the Taylor series expansion of the function in powers of $z - z_0$ takes the form*

$$f(z) = a_n(z - z_0)^n + a_{n+1}(z - z_0)^{n+1} + a_{n+2}(z - z_0)^{n+2} + \ldots,$$

i.e. if $a_0 = a_1 = a_2 = \ldots = a_{n-1} = 0$, $a_n \neq 0$, then f is said to have a zero of order n at z_0. In the particular case $f(z) = a_1(z - z_0) + a_2(z - z_0)^2 + \ldots$, the function is said to have a simple zero, *or zero of order 1, at z_0.*

Evidently in the case of a zero of order n, all the terms in the expansion have a common factor $(z - z_0)^n$, and we may write

$$f(z) = (z - z_0)^n \phi(z),$$

where ϕ is a regular function of z [since the radius of convergence of $a_n + a_{n+1}(z - z_0) + \ldots$ is exactly the same as that of $a_n(z - z_0)^n + a_{n+1}(z - z_0)^{n+1} + \ldots$] which does not vanish at z_0 [since $\phi(z_0) = a_n \neq 0$.]

12.2. Theorem

The zeros of a regular function are isolated.

By this we mean that if z_0 is a zero of f, then there exists a neighbourhood of z_0 in which z_0 itself is the only zero.

Proof. Let us suppose that the order of the zero z_0 is n and write $f(z) = (z-z_0)^n \phi(z)$, where ϕ is regular in a neighbourhood of z_0, and that $|\phi(z_0)| = c > 0$.

Then, since ϕ is continuous at z_0, there exists a neighbourhood of z_0 within which $|\phi(z)-\phi(z_0)| < \frac{1}{2}c$. Let this be the neighbourhood $|z-z_0| < \delta$. Within this neighbourhood we have

$$|\phi(z)| = |\phi(z_0)-[\phi(z_0)-\phi(z)]| \geqslant |\phi(z_0)|-|\phi(z_0)-\phi(z)|$$
$$> c-\tfrac{1}{2}c.$$
$$= \tfrac{1}{2}c.$$

Thus in this neighbourhood $\phi(z)$ does not vanish; nor, evidently does $(z-z_0)$, except at z_0 itself. Hence $f(z)$ does not vanish in this neighbourhood except at z_0.

We may express this result by the statement that $f(z)$ does not vanish in the *deleted neighbourhood* of z_0, by which we mean the neighbourhood of z_0 excluding z_0 itself.

12.3. Singular Points

It may happen that there are certain points z within a domain S at which either $f(z)$ is not defined or f is not regular. Such points are called *singular points* or *singularities*. A singularity which has nevertheless a deleted neighbourhood in which f is regular is called an *isolated singularity*. We proceed to consider three types of isolated singularity.

(a) *Removable singularities*

Let f be regular in a domain S except at the point z_0. Let $f(z_0) = a$, and suppose $\lim_{z \to z_0}$ exists, and equals b. We define a new function g on S by $g(z) = f(z)$ $(z \neq z_0)$, $g(z_0) = b$. Then the new function g is regular over S. (It is evidently continuous at z_0 by construction, and elsewhere since it is the same as f, and the regularity follows easily from Morera's theorem. See Exercise 12.3.1.)

The replacement of f by g having removed the singularity, we refer to such a singularity as *removable*.

(b) *Poles*

Suppose that in some neighbourhood of a point z_0, f has a Laurent expansion of the form

$$f(z) = \sum_{n=0}^{\infty} a_n(z-z_0)^n + \sum_{n=1}^{m} b_n(z-z_0)^{-n}.$$

That is, assume that only a finite number of the b_r are non-zero, and that b_m is not zero but that all b_r for $r > m$ are zero. Then z_0 is said to be a *pole*, of order m, of f. When $m = 1$, the pole is said to be simple. Note that $f(z)$ is not defined for $z = z_0$ by the above expression, if at all.

A function of the form $f(z) = \phi(z)/(z-z_0)^k$ ($z \neq z_0$), where k is a positive integer and ϕ is regular in the domain considered, will have a pole of order k at z_0 if $(z-z_0)$ is not a factor of $\phi(z)$. For, $\phi(z)$ may be expanded as a series of positive powers of $(z-z_0)$, starting with a non-zero constant term [otherwise $(z-z_0)$ divides $\phi(z)$], and division of this series term by term by $(z-z_0)^k$ will convert the expression to the form indicating a pole of order k.

We may in fact go further. We shall find the following theorem of great importance in due course.

THEOREM. *If $f(z) = p(z)/q(z)$, where p and q are both regular at z_0, and if $p(z) \neq 0$, then f has a pole of order m at z_0 if and only if q has a zero of order m at z_0.*

Proof. From the previous paragraph it is clear that if z_0 is a zero of q, then it is a pole (of the same order) of f.

On the other hand, if there is a pole of order k at z_0, then we have

$$f(z) = \phi(z) + b_1/(z-z_0) + \ldots + b_k/(z-z_0)^k,$$

where ϕ is regular at z_0 [and $\phi(z)$ consists of a series of non-negative powers of $z-z_0$]. Bringing these terms to a common denominator gives an expression of the form $f(z) = F(z)/(z-z_0)^k$, where F is regular at z_0 and contains no factor $(z-z_0)$. Thus $p(z)/q(z) = F(z)/(z-z_0)^k$, and so $q(z) = (z-z_0)^k \psi(z)$, where $\psi(z) = p(z)/F(z)$.

However, ψ is regular at z_0 since $F(z_0) \neq 0$ (see Exercise 4.6.2) and does not vanish at z_0, since $p(z_0) \neq 0$. Hence q has a zero of order k at z_0.

Poles, in common with zeros, are isolated. (See Exercise 12.3.3 for the proof of this important result.)

(c) *Essential singularities*

When the principal part of a Laurent expansion of a function f in the neighbourhood of a point z_0, i.e the part containing negative powers of $(z - z_0)$, contains an infinite number of terms, the function is said to have an (isolated) *essential singularity* at z_0. As examples, the functions $e^{1/z}$, $\sin (1/z)$ both have essential singularities at $z = 0$, since

$$e^{1/z} = 1 + z^{-1} + z^{-2}/2! + \ldots + z^{-n}/n! + \ldots,$$
$$\sin (1/z) = z^{-1} - z^{-3}/3! + z^{-5}/5! - \ldots$$

Non-isolated essential singularities, where a Laurent expansion in the neighbourhood of the singularity is not possible, will not be considered in this book. In what follows, we shall for the most part be concerned with poles rather than essential singularities.

A function whose only singularities in a domain D are poles, is said to be *meromorphic* in D.

12.4. The Integration of $b_k/(z - z_0)^k$

In the following sections we shall frequently have to consider term-by-term integration of a Laurent expansion of a function. It will be convenient to examine at this point the integration of a single term of the expansion, the path of integration being a circle with centre z_0 and radius r, the function being considered regular at all points on and within the circle except at z_0. Denoting this contour by C we have to evaluate

$$\int_C \frac{b_k}{(z - z_0)^k} \, dz.$$

Let us replace z by $z_0 + re^{i\theta} = z_0 + r\cos\theta + ir\sin\theta$, since z lies on C. Then we have

$$dz/d\theta = -r\sin\theta + ir\cos\theta$$
$$= i(ir\sin\theta + r\cos\theta)$$
$$= ir(\cos\theta + i\sin\theta)$$
$$= ire^{i\theta},$$

and as θ varies from 0 to 2π the point z goes once around C. The integral thus reduces to

$$\int_0^{2\pi} \frac{b_k ire^{i\theta}}{r^k e^{ki\theta}} d\theta \quad \text{or to} \quad \int_0^{2\pi} ib_k r^{1-k} e^{(1-k)i\theta} d\theta.$$

There are two possibilities to consider; namely, $k = 1$ and $k \neq 1$. Taking the latter first, the integral is simply

$$\frac{ib_k r^{1-k}}{i(1-k)} \left[e^{(1-k)i\theta} \right]_0^{2\pi},$$

which reduces to zero since if m is an integer then $e^{2m\pi i} = e^0 = 1$.

On the other hand, if $k = 1$ we have just to consider the integral $\int_0^{2\pi} ib_1 \, d\theta$ to which the general form reduces, and this is simply $2\pi ib_1$.

Now in the Laurent expansion of a function f the (possibly infinite) series of positive powers of $(z - z_0)$ gives a regular function which, by Cauchy's theorem, gives 0 when integrated round the contour C. It follows, therefore, from all this, that if we integrate the Laurent series for f term by term around the contour C (where r is chosen such that the Laurent series converges), the only term which survives is the b_1 term, which appears finally as $2\pi ib_1$. For this reason the coefficient b_1 is called the *residue* of the function (represented by the series) at z_0.

12.5. The Evaluation of Residues

We will restate more clearly the above definition.

DEFINITION. *The residue at z_0 of a function f which has an isolated singularity at z_0 is the coefficient of $(z-z_0)^{-1}$ when $f(z)$ is expanded as a Laurent series of positive and negative powers of $(z-z_0)$.*

We consider the case of a simple pole, and of a multiple pole, separately.

(a) *Simple poles*

In a deleted neighbourhood of the simple pole z_0 we have

$$f(z) = \phi(z)+b_1/(z-z_0),$$

where ϕ is regular, and $\phi(z)$ is in fact a series of ascending powers of $(z-z_0)$ possibly preceded by a constant. We wish to determine b_1. Clearly for all values of z in the deleted neighbourhood of z_0 we may write

$$f(z)(z-z_0) = \phi(z)(z-z_0)+b_1.$$

If we now allow z to approach z_0 the first term on the right approaches $\phi(z_0)\times 0 = 0$, and we are left with the result

$$b_1 = \lim_{z \to z_0} (z-z_0)f(z).$$

If, as often happens, the function f is of the form $f(z) = (z-z_0)^{-1}g(z)$, where g is continuous at z_0, then there is no problem:

$$b_1 = \lim_{z \to z_0} f(z)(z-z_0)$$

obviously gives at once $b_1 = g(z_0)$.

In a more complicated situation, such as $f(z) = 1/(z^4+a^4)$, where z_0 is one of the four values of z that make the denominator vanish (viz. $z = ae^{(2n+1)i\pi/4}$, $n = 0, 1, 2, 3$), giving

$$(z-z_0)f(z) = (z-z_0)/(z^4+a^4)$$

as the expression of which the limit is required, rather than factorize the denominator it is helpful to make use of *l'Hôpital's rule* (Exercise 12.5.2), which may be stated in the following form:

If f and g are regular functions in a domain containing the point z_0, and if $f(z_0) = g(z_0) = 0$, $g'(z_0) \neq 0$, then

$$\lim_{z \to z_0} f(z)/g(z) = f'(z)/g'(z_0).$$

By applying this in the example given, the residue of f at z_0 is easily seen to reduce to $1/4z_0^3$, which may be at once evaluated on substituting the particular value of z_0 involved.

Before leaving the subject of the evaluation of the residue at a simple pole, we note another useful result which frequently simplifies the calculation when applying l'Hôpital's rule. Suppose that f has a simple pole at z_0 and that $f(z)$ is given in the form $\phi(z)/\psi(z)$. The residue of f at z_0 is

$$\lim_{z \to z_0} (z - z_0)f(z) = \lim_{z \to z_0} (z - z_0)\phi(z)/\psi(z).$$

If we are using l'Hôpital's rule we need to evaluate

$$d/dz[(z - z_0)\phi(z)]/\psi'(z_0)$$

at z_0. However it is easily verified that the value of $d/dz[(z - z_0)\phi(z)]$ at z_0 is just $\phi(z_0)$. Therefore the residue is $\phi(z_0)/\psi'(z_0)$. It is worth realizing this fact that the calculation of ϕ' may not be necessary, since the function ϕ may well be one which is not easy to differentiate.

As an example, let us evaluate the residue at $z = i$ of the function f given by $f(z) = e^{z\sqrt{(1 + 2z^2)}}/(z^4 - 1)$. Here $\phi(z) = e^{z\sqrt{(1 + 2z^2)}}$, $\psi(z) = z^4 - 1$. We apply l'Hôpital's rule to the expression $(z - i)\phi(z)/\psi(z)$ with ϕ and ψ as stated. By the above we may read off the residue at i at once in the form

$$b_1 = \phi(i)/\psi'(i)$$
$$= e^{i\sqrt{(1-2)}}/4i^3$$
$$= e^{-1}/-4i \quad \text{or} \quad e/-4i$$

according to whichever value of the square root we are taking to define f near the point i,

$$= i/4e \quad \text{or} \quad ie/4$$

as the case may be.

(b) *Double poles. First method*

If $f(z) = \phi(z)+b_1/(z-z_0)+b_2/(z-z_0)^2$, where ϕ is regular in a neighbourhood of z_0, the method of procedure we used above in order to evaluate b_1 will obviously no longer apply since the last term will cause trouble as we proceed to the limit. If, instead, we multiply through by $(z-z_0)^2$, we obtain

$$(z-z_0)^2 f(z) = (z-z_0)^2 \phi(z)+(z-z_0)b_1+b_2,$$

isolating the wrong constant. However, if we now differentiate with respect to z, we have

$$d/dz\{(z-z_0)^2 f(z)\} = (z-z_0)^2 \phi'(z)+2(z-z_0)\phi(z)+b_1,$$

and we now see, on letting z approach z_0, that

$$b_1 = \lim_{z \to z_0} d/dz\{(z-z_0)^2 f(z)\}.$$

(c) *Double poles. Second method*

Whether the above procedure is convenient evidently depends on the nature of the expression in brackets which has to be differentiated, possibly after simplification. Occasionally it is more convenient to carry out a direct expansion of $f(z)$ in positive and negative powers of $z - z_0$, sufficient to enable b_1 to be read off. The technique is to write $z-z_0 = t$, so that $z = z_0+t$, and to expand $f(z_0+t)$ in powers of t by any convenient method (we know the expansion will be unique), finally selecting the coefficient of t^{-1}, which will be by definition the residue b_1.

As an example, consider the problem of finding the residue at ib of the function given by the expression $e^{z^2}/(z^2+a^2)(z^2+b^2)^2$. (We notice in passing that as the denominator can be factorized further to $(z+ia)(z-ia)(z+ib)^2(z-ib)^2$, and since none of these factors divides the numerator, the function has single poles at $-ia$, $+ia$, and double poles at $-ib$, $+ib$.) For comparison of methods, the differentiation method outlined above will also be used on the same expression afterwards.

The first step is to replace z by $ib+t$, leading to

$$e^{(ib+t)^2}/(a^2-b^2+2ibt+t^2)(2ibt+t^2)^2.$$

We now observe that the expansion of the various factors is best accomplished with the aid of the familiar exponential and binomial expansions; and to this end we reorganize the expression in the form

$$e^{-b^2}e^{2ibt}e^{t^2}\Big/(a^2-b^2)\left(1+\frac{2ibt+t^2}{a^2-b^2}\right)(-4b^2t^2)(1+t/2ib)^2.$$

Observing the factor t^2 in the denominator, we see that we need only look for the coefficient of t in the expansion of

$$\left[e^{-b^2}e^{2ibt}e^{t^2}/4b^2(b^2-a^2)\right]\times\left(1+\frac{2ibt+t^2}{a^2-b^2}\right)^{-1}\times(1+t/2ib)^{-2}.$$

Note now that terms involving t^2 or higher powers of t will contribute nothing to the term in which we are interested. Therefore, since $e^{t^2}=1+t^2+t^4/2!+\ldots$ we may forget about the e^{t^2} factor. Expanding the remaining factors, and neglecting terms involving t^2 and higher powers of t, we obtain the following expression from which we need to extract the coefficient of t:

$$\frac{e^{-b^2}}{4b^2(b^2-a^2)}\times(1+2ibt)\left(1-\frac{2ibt}{a^2-b^2}\right)\times(1-t/ib).$$

On expanding this expression we have no difficulty in reading off the coefficient of t as

$$b_1=\frac{-e^{-b^2}}{4b^2(b^2-a^2)}\left[2ib-\frac{2ib}{a^2-b^2}-\frac{1}{ib}\right],$$

which may be simplified slightly to give

$$b_1=\frac{-ie^{-b^2}}{4b^3(b^2-a^2)^2}\times[2b^2(a^2-b^2)+a^2-3b^2].$$

For the alternative method we need to evaluate the limit as z approaches ib of

$$d/dz[(z-ib)^2 e^{z^2}/(z^2+a^2)(z^2+b^2)^2].$$

The expression to be differentiated can be simplified slightly, so that we need only evaluate the limit of

$$d/dz[e^{z^2}/(z^2+a^2)(z+ib)^2].$$

The differentiation leads to

$$\frac{\{(z^2+a^2)(z+ib)^2 2ze^{z^2} - e^{z^2}(z^2+a^2)2(z+ib) + 2z(z+ib)^2\}}{(z^2+a^2)^2(z+ib)^4}.$$

In order to evaluate the limit we seek, it is only necessary to substitute $z = ib$ in this expression, since the denominator does not now vanish at $z = ib$. We obtain

$$e^{-b^2}\{(a^2-b^2)(-4b^2)(2ib) - 2(a^2-b^2)(2ib) - 2ib(-4b^2)\}/(a^2-b^2)^2 16b^4$$

which easily reduces to the same result as before.

(d) *Multiple poles*

These may be treated similarly to double poles. Evidently the expansion procedure is still valid and presents no new type of difficulty if the pole is of higher order than 2. On the other hand, the differentiation method will involve, for a pole of order m, a derivative of order $m-1$. The appropriate formula, the proof of which is left as an exercise (Exercise 12.5.3), is

$$b_1 = [1/(m-1)!] \lim_{z \to z_0} d^{m-1}/dz^{m-1}[(z-z_0)^m f(z)].$$

It should be noted that the residues at simple poles may be obtained by the expansion method quite easily, although not usually so quickly as by the standard method given in (a) above. For $m = 1$ the differentiation method reduces to the standard method.

12.6. Behaviour of a Function near a Pole

While the value of a function is not defined at a pole, it is important to notice that, whatever the order of the pole, $|f(z)|$ approaches infinity as z approaches the pole. For, let $f(z) = (z-z_0)^{-k}\phi(z)$, where ϕ is regular and $\phi(z_0) = a \neq 0$. Then by continuity of ϕ at z_0, an $r > 0$ exists such that $|\phi(z)| > \frac{1}{2}|a|$ for $|z-z_0| < r$, and so $|f(z)| > \frac{1}{2}|a| \, |z-z_0|^{-k}$, from which the result follows at once.

12.7. Singularities at Infinity

Let f be any complex function defined on the whole plane, or at least outside some circle of radius R. In order to examine the behaviour of $f(z)$ as z approaches infinity, it is convenient to replace z by $1/w$ and to consider the behaviour of $f(1/w)$ as w approaches zero. If $f(1/w)$ has a pole of order m at 0, with principal part $b_1/w+b_2/w^2+ \ldots +b_m/w^m$, then we say that f has a pole at *infinity*, with principal part $b_1z+b_2z^2+ \ldots +b_mz^m$. Similar definitions apply for *essential singularities at infinity*, and for functions *regular at infinity*.

Exercises

12.1.1. Determine the orders of the zeros (at the points stated) of the functions given by the following expressions:

(a) $z^2 \sin z$ at the origin;

(b) $z^2(\pi/2-z) \cos z$ at $\pi/2$;

(c) $\sin^2 3z$ at the origin.

12.1.2. Determine all the zeros of the function f defined by:

(a) $f(z) = \sinh az$ (a real);

(b) $f(z) = (z^3+a^3)^2$ (a real);

(c) $f(z) = z^2+z+1$;

(d) $f(z) = z^4+2z^2+1$.

12.2.1. If $f(z) = (z+i)^2(z-1)$, and we write $\phi(z)$ for $(z-1)$, verify the following:

(a) f has a double zero at $z = -i$;

(b) $|\phi(-i)| = \sqrt{2}$;

(c) $|\phi(z)-\phi(-i)| < \frac{1}{2}\sqrt{2}$ provided that $|z+i| < \frac{1}{2}\sqrt{2}$, i.e. $z = re^{i\theta}-i$, where $r < \frac{1}{2}\sqrt{2}$;

(d) for values of z given by (c) the minimum value of $|\phi(z)|$ is greater than $\frac{1}{2}\sqrt{2}$.

12.2.2. Let f be regular in a domain D, and let R be a rectangle lying inside D. Prove that the number of zeros of f which lie inside R is finite. (Use the Bolzano–Weierstrass property.)

12.2.3. Verify that $\sin(1/z)$ has an infinite number of zeros lying within any circle with centre 0, and is hence not regular at the origin.

12.2.4. Give an example of a function which is regular over the entire complex plane and has an infinite number of zeros.

12.3.1. Let f be regular in a domain D, except perhaps at a single point P, and let f be continuous at P. Show that:

(a) if C is any closed contour in D not containing P as an interior or boundary point, then $\int_C f(z)\,dz = 0$;

(b) if P lies within the interior of C, then $\int_C f(z)\,dz = 0$ (use a cross-cut);

(c) if P lies on C, $\int_C f(z)\,dz = 0$. (Hint: consider a small circular arc excluding P from the contour.) Deduce that f is regular everywhere within D.

12.3.2. Identify and remove the removable singularities from the functions given by the following expressions:

(a) $z^2/\sin 2z$; (c) $\tan z/\sinh z$;
(b) $(z+ia)/(z^2+a^2)$; (d) $\sin(z+\pi i)/(z^4-\pi^4)$.

12.3.3. By considering the Laurent expansion, valid for points in the annulus between the circles $|z-z_0| = r_1$, $|z-z_0| = r_2$ $(r_1 < r_2)$, of a function f with a pole at z_0, deduce that there are no other singularities within the smaller circle, and thus that poles are isolated singularities.

12.3.4. Show that the number of poles of a meromorphic function which lie within any simple closed contour is finite (cf. Exercise 12.2.2).

12.3.5. Identify the poles of the functions given by the following expressions, and state the order of each pole;

(a) $(z^2+4z-5)/(z^3+1)^2(z^2+1)^3$; (d) $\tan z$;
(b) $(z^2+2)/\sin z$; (e) $\tanh 2z$;
(c) $1/z^2 \sin 3z$; (f) $\tanh^2 z$.

12.3.6. Identify the nature of the singularities at all the singular points of the following functions of z:

(a) $(z-2)/(z^2-4)$; (d) $z\sin(1/z)$;
(b) $(z-2)/(z^2+4)$; (e) $z^n e^{1/z}$;
(c) $z/\sin z$; (f) $e^{az}\operatorname{sech} bz$.

12.4.1. Evaluate from first principles the integral round the unit circle $|z| = 1$ of each of the following expressions:

(a) $1/z$;   (c) $1/z^3$;

(b) $1/z^2$;   (d) $(az^2+bz+c)/z$.

Also integrate each expression round the upper semicircle, from amp $z = 0$ to amp $z = \pi$, and the lower semicircle from amp $z = \pi$ to amp $z = 2\pi$.

12.5.1. Evaluate the residues of the following functions, by each of two methods where practicable, at each pole:

(a) $1/(z^2+4)$;   (g) $(2z+1)/(z-\tfrac{1}{2})$;   (m) $1/(a^2+z^2)^2$;

(b) $1/(az^2+bz+c)$;   (h) $(z^2+1)/(z^3+1)$;   (n) $1/(z+3i)^2$;

(c) $(z-2)/(z+2)$;   (i) $(2z+1)/\sin 2z$;   (o) $e^{iaz}/(z^2+a^2)/(z^2+b^2)$;

(d) $\operatorname{cosec} z$;   (j) $e^{iz}/(z^2+1)$;   (p) $e^{3z^3+2z+1}/(z^4+16)$;

(e) $(2z+1)/(2z-1)$;   (k) $z^2/(1+z^4)$;   (q) $(z^2+1)/(z+2)^3$;

(f) $(z+\tfrac{1}{2})/(2z-1)$;   (l) $1/z^3(a^2+z^2)$;   (r) $\sin z/(z^2+1)^3$.

12.5.2. If f and g are differentiable functions at z_0, and if $g'(z_0) \neq 0$, prove, by writing

$$\frac{f(z)-f(z_0)}{g(z)-g(z_0)} = \frac{[f(z)-f(z_0)]/(z-z_0)}{[g(z)-g(z_0)]/(z-z_0)} \qquad (z \neq z_0),$$

that

$$\lim_{z \to z_0} [f(z)-f(z_0)]/[g(z)-g(z_0)] = f'(z_0)/g'(z_0).$$

State the special case which arises when $f(z_0) = g(z_0) = 0$.

12.5.3. Prove the formula of § 12.5(d) for the residue at a pole of order m.

12.6.1. If $f(z) = 1-2z+1/z^2$, determine a suitable r (which need not be the greatest possible) such that if $|z| < r$, then $|f(z)| > 100$.

12.6.2. Show that $(\sin z)/z^2$ approaches infinity as z approaches zero, while $(\sin 2z)/z$ does not. Comment.

12.6.3. Show that there are points z arbitrarily near the origin where $e^{1/z} = i, -i,$ $1, -1, e$. (Hint: consider $e^{w+2n\pi i}$ for large n.)

Does $e^{1/z}$ have a pole at the origin? State a value which is not taken by $e^{1/z}$ at any point near the origin (or elsewhere).

12.7.1. Discuss the behaviour at infinity of the functions given by the following expressions:

(a) z^2;   (d) $e^{1/z}$;   (g) $z^2(z+a)/(z^4+b^4)$;

(b) $(z+1)/(z-1)$;   (e) $\cosh 1/z$;   (h) $(z^3+a^3)/(z^2+a^2)$;

(c) $\sin z$;   (f) $1/z^n (n = 1, 2, \ldots)$;   (i) $z^n \sin 1/z$.

CHAPTER 13

CONTOUR INTEGRATION

13.1. Cauchy's Residue Theorem

We now come to a theorem which is of great importance in complex variable theory, and has many applications.

THEOREM (Cauchy's residue theorem). *Let C be a simple closed contour, and let f be regular on C and meromorphic within C. Then*

$$\int_C f(z)\, dz = 2\pi i \sum \mathcal{R},$$

where $\sum \mathcal{R}$ denotes the sum of the residues at those poles of f which lie within C.

Proof. Let the poles occur at $z_1, z_2, \ldots, z_p$. Since poles are isolated we may construct circles $C_1, C_2, \ldots, C_p$ with these points as centres and with radii sufficiently small to ensure that each circle lies inside C and contains only one pole (at its centre). By means of cross-cuts (Fig. 13.1, and see also § 10.4) we can construct a contour within and on which f is regular, since the poles are now exterior to the new contour, and by the application of Cauchy's theorem (§ 10.2 *et seq.*) we easily obtain the result

$$\int_C f(z)\, dz - \int_{C_1} f(z)\, dz - \ldots - \int_{C_p} f(z)\, dz = 0$$

or

$$\int_C f(z)\, dz = \sum_{r=1}^{p} \int_{C_r} f(z)\, dz.$$

205

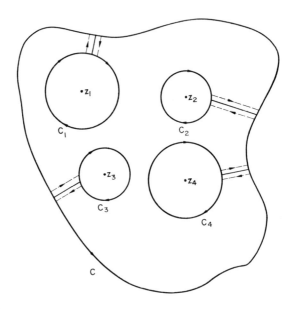

FIG. 13.1. Constructing a simple closed contour excluding all poles.

Now in the neighbourhood of the pole z_r we may write $f(z)$ in the form

$$f(z) = \phi_r(z) + b_{r,1}/(z - z_r) + b_{r,2}/(z - z_r)^2 + \ldots,$$

where ϕ_r is regular and the principal part of the expansion contains only a finite number of terms. On integrating term by term we see, from Cauchy's theorem and § 12.4, that

$$\int_{C_r} f(z)\, dz = 2\pi i b_{r,1},$$

with similar results for the integrals round the other poles. The result follows.

Since we have already investigated the problem of calculating residues at poles, it will be appreciated that the above theorem provides a simple and powerful way of evaluating certain integrals. In later sections we shall show how one can evaluate certain types of real integral by means of this method when other standard methods present difficulties.

13.2. Periodic Functions

We first consider integrals of the form

$$I = \int_0^{2\pi} F(\sin\theta, \cos\theta)\,d\theta.$$

By means of the substitution $e^{i\theta} = z$, we may reduce this to a contour integration round the unit circle; for we have

$$\cos\theta = \tfrac{1}{2}(z+1/z), \quad \sin\theta = \frac{1}{2i}(z-1/z),$$

and we easily see that $d\theta/dz = 1/iz$ (from $z = \cos\theta + i\sin\theta$). Moreover, as θ runs from 0 to 2π, the point $z = e^{i\theta}$ traverses the unit circle once in the counter-clockwise direction.

If the function $F(\cos\theta, \sin\theta)$ is a rational function of $\sin\theta$ and $\cos\theta$, i.e. if it is of form $p(\theta)/q(\theta)$, where p and q are polynomials in $\sin\theta$ and/or $\cos\theta$, and if F is defined everywhere in the range of integration [$q(\theta)$ does not vanish, for example], then the integral reduces to the form

$$I = \int_C f(z)\,dz,$$

where f is a rational function of z which has no singularities on the unit circle C. We can therefore immediately evaluate the integral as

$$I = 2\pi i \sum \mathcal{R},$$

where $\sum \mathcal{R}$ denotes the sum of the residues at those poles of f which lie within the unit circle. As an example (which may in fact easily be integrated by conventional methods), consider

$$I = \int_0^{2\pi} d\theta/(1+a\cos\theta) \quad (-1 < a < 1).$$

(The reason for the restriction on a is to prevent the integrand becoming infinite on the range of integration.)

On transforming the integral as above we find that it easily reduces to

$$I = \int_C \frac{2}{i(2z + az^2 + a)} \, dz$$

or, on rearranging, to

$$I = \int_C \frac{2}{ia(z^2 + 2z/a + 1)} \, dz.$$

In order to find the poles (there are two of them, which we denote by z_1 and z_2) we solve the quadratic equation $z^2 + 2z/a + 1 = 0$, noting incidentally that since the product of the roots is 1 and thus the product of their moduli is 1 also, only one of the roots can lie in the interior of the unit circle C. Let this root be denoted by z_1. If for the time being we rewrite the integral in the form

$$I = \int_C \frac{2}{ia(z - z_1)(z - z_2)} \, dz,$$

we easily read off the residue at the (simple) pole z_1 as $2/ia(z_1 - z_2)$. (See § 12.5.) Now solving the quadratic equation gives

$$z_1 = [-1 + \sqrt{(1 - a^2)}]/a, \qquad z_2 = [-1 - \sqrt{(1 - a^2)}]/a,$$

and the application of the residue theorem thus gives

$$\begin{aligned}
I &= 2\pi i \times 2/ia(z_1 - z_2) \\
&= 2\pi i \times 2/\{ia[2\sqrt{(1 - a^2)}]/a\} \\
&= 2\pi/\sqrt{(1 - a^2)}.
\end{aligned}$$

13.3. Infinite Integrals

We now apply the residue theorem to the problem of evaluating certain types of infinite integral. Consider a function f which is regular along the real axis and meromorphic in the upper half-plane.

Let C_R denote the closed contour consisting of the line segment C_R' from $-R$ to $+R$ ($R > 0$) on the real axis, followed by the semicircular arc C_R'' of the circle $|z| = R$ from $(R, 0)$ to $(-R, 0)$ in the upper half-plane. (Fig. 13.2.)

Providing that none of the poles of f lie on the semicircle C_R'' we can clearly use the residue theorem to evaluate $\int\limits_{C_R} f(z)\ dz$. Now it may happen in particular cases that as R approaches infinity the contribu-

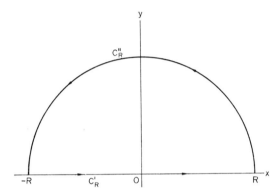

FIG. 13.2. Integration round the upper semicircle of $|z| = R$.

tion to the integral from the semicircle C_R'' approaches zero. If this is so, then since

$$\int\limits_{C_R} f(z)\ dz = \int\limits_{C'} f(z)\ dz + \int\limits_{C''} f(z)\ dz$$

we have

$$\lim_{R \to \infty} \int\limits_{C_R} f(z)\ dz = \lim_{R \to \infty} \int\limits_{C_R'} f(z)\ dz.$$

If, further, we assume that there are a finite number of poles of f in the upper half-plane (and this will certainly be the case—cf. Exercise 12.3.4—provided that $f(z)$ approaches zero as R approaches infinity) we may assume R to be so large that C_R contains all these poles. In this case we know from the residue theorem that the integral $\int\limits_{C_R} f(z)\ dz$

does not depend on R, and is simply equal to $2\pi i \sum \mathcal{R}$, where $\sum \mathcal{R}$ is the sum of the residues in the upper half-plane. But

$$\lim_{R \to \infty} \int_{C'} f(z)\, dz$$

is exactly what is meant by

$$\int_{-\infty}^{\infty} f(x)\, dx.$$

Therefore we have proved

$$\int_{-\infty}^{\infty} f(x)\, dx = 2\pi i \sum \mathcal{R}.$$

We will again take as our example an integral which may easily be evaluated by normal methods—in fact a standard integral. (Obviously a simple example is preferable for a first demonstration; some more formidable examples will follow.)

We consider

$$\int_{-\infty}^{\infty} \frac{dx}{a^2 + x^2} \quad (a > 0).$$

Here we take f to be given by $f(z) = 1/(a^2 + z^2)$. Write I_1 for the integral of f along C_R''. Since

$$|a^2 + z^2| = |z^2 + a^2|$$
$$\geqslant |z^2| - |a^2| \quad (\S\ 1.14)$$
$$= R^2 - a^2 \quad \text{on } C_R''$$

we have $1/|a^2 + z^2| \leqslant 1/(R^2 - a^2)$ on C_R'', and so, from § 9.5,

$$|I_1| \leqslant [1/(R^2 - a^2)]\pi R,$$

where we have assumed $R > a$.

We see that I_1 will approach zero as R approaches infinity, so this is a case in which the method described above can be applied. Now in the upper half-plane there is one simple pole of f, at $z = ia$. The residue there is easily seen to be $1/2ia$, and, on multiplying by $2\pi i$, we finally obtain

$$\int_{-\infty}^{\infty} \frac{dx}{x^2+a^2} = \frac{\pi}{a}.$$

[This result may be easily verified from the fact that the indefinite integral is here known to be $(1/a)\tan^{-1}(x/a)$.]

As a harder example consider

$$I = \int_{0}^{\infty} \frac{x^2}{(x^4+a^4)^2}\, dx \quad (a > 0).$$

Although the integral this time runs from 0 to ∞, this gives no trouble since the function is "even" $[f(x) = f(-x)]$, so we merely need to halve the integral from $-\infty$ to $= +\infty$ along the complete real axis.

Considering the function f given by the expression $z^2/(z^4+a^4)^2$ we note that the integral of f along the semicircular arc C_R'' tends to zero as R tends to infinity, being less than

$$\frac{R^2}{(R^4-a^4)^2}\, \pi R$$

in modulus, which is certainly less than $\pi R^3/(\tfrac{1}{2}R^4)^2$ if R is large enough. We also note that there are two double poles in the upper half-plane, since poles occur where

$$z^4 = -a^4 = a^4 e^{i\pi + 2n\pi i} \quad (n = 0, 1, 2, \ldots)$$

giving

$$z = a e^{(2n+1)i\pi/4} \quad (n = 0, 1, 2, 3).$$

Thus the poles at which we need the residues occur at $z = ae^{i\pi/4}$ and $z = ae^{3i\pi/4}$. Let us denote either of these poles, for the moment, by z_1. We now substitute $z = z_1 + t$ and extract the coefficient of t^{-1}. We have

$$f(z_1+t) = (z_1+t)^2/(4z_1^3 t + 6z_1^2 t^2 + 4z_1 t^3 + t^4)^2.$$

Notice that we have a factor t^2 in the denominator; if we extract that, we need the coefficient of t in the remainder of the expression. We hence neglect higher powers of t than the first, and seek merely the coefficient of t in the expansion of

$$(z_1^2 + 2z_1 t)(4z_1^3)^{-2}(1 + 3t/2z_1)^{-2},$$

i.e. the coefficient of t in the expansion of

$$(1/16z_1^6)(z_1^2 + 2z_1 t)(1 - 3t/z_1).$$

This is evidently $-1/16z_1^5$, since

$$(z_1^2 + 2z_1 t)(1 - 3t/z_1) = z_1^2 - z_1 t - 6t^2.$$

Now of the two poles, consider first $z_1 = ae^{i\pi/4}$. In this case $-1/16z_1^5$ reduces to

$$-e^{-5i\pi/4}/16a^5 = (-\cos 5\pi/4 + i \sin 5\pi/4)/16a^5$$
$$= (1/\sqrt{2} - i/\sqrt{2})/16a^5.$$

Similarly, if $z_1 = ae^{3i\pi/4}$, the expression $-1/16z_1^5$ reduces to

$$(-\cos 15\pi/4 + i \sin 15\pi/4)/16a^5 = (-1/\sqrt{2} - i/\sqrt{2})/16a^5,$$

whence the sum of the two residues is $-i\sqrt{2}/16a^5$, and the value of the integral is thus

$$\tfrac{1}{2} \times 2\pi i(-i\sqrt{2})/16a^5 = \pi\sqrt{2}/16a^5.$$

13.4. Infinite Integrals involving Trigonometric Functions

The next type of integral we consider is one of the form

$$I = \int_{-\infty}^{\infty} \cos mx f(x)\, dx \quad (m > 0)$$

or a similar expression containing $\sin mx$, where f is a rational function (with real coefficients) with the denominator of higher degree than the numerator. In considering the behaviour of this expression on the upper semicircle C_R'', we note that $\cos mz = \frac{1}{2}(e^{imz} + e^{-imz})$, and if

$$z = Re^{i\theta} = R(\cos \theta + i \sin \theta)$$

then the e^{imz} term may be written $e^{imR \cos \theta - mR \sin \theta}$; its modulus $e^{-mR \sin \theta}$ evidently approaches zero as R approaches infinity, for $0 < \theta < \pi$, and has the value 1 for $\theta = 0$ or π. Thus if $f(z)$ approaches zero as z approaches infinity in the upper half-plane, so does $e^{iz}f(z)$.

The e^{-imz} term, on the other hand, has a modulus of $e^{mR \sin \theta}$ (verify this), which for $0 < \theta < \pi$ approaches infinity with R. In fact if f is a rational function, with $f(z)$ of form $p(z)/q(z)$, the ratio $e^{mR \sin \theta}/q(Re^{i\theta})$ will approach infinity with R, whatever the degree of q.

A similar difficulty arises with $\sin mz$.

Fortunately we are able to escape from these difficulties in the following way. We first note that since f takes real values along the real axis the integral

$$\int_{-\infty}^{\infty} \cos mx f(x) \, dx$$

is just the real part of

$$\int_{-\infty}^{\infty} e^{imx} f(x) \, dx.$$

As we have seen above, $e^{imz}f(z)$ approaches zero as z approaches infinity in the upper half-plane whenever $f(z)$ does so. If

$$\int_{C_R''} e^{imz} f(z) \, dz$$

also approaches zero as R approaches infinity, we can use the method of § 13.3 to evaluate

$$\int_{-\infty}^{\infty} e^{imx} f(x) \, dx;$$

taking the real part then gives the integral originally required. Of course,

$$i \int_{-\infty}^{\infty} \sin mx f(x) \, dx$$

is obtained by taking the imaginary part.

Again we consider an example. Let us take

$$I = \int_{-\infty}^{\infty} \frac{\cos x}{(x^2+4)(x^2+1)}\, dx.$$

We proceed to examine $\displaystyle\int_{C_R} \frac{e^{iz}}{(z^2+4)(z^2+1)}\, dz$, where C_R is as usual the contour consisting of a part C_R' of the real axis and the upper semicircle C_R'' associated with it. Evidently for $|z| > R$, the modulus of the integrand is less than $1/(R^2-4)(R^2-1) < 1/\tfrac{1}{2}\,R^4$ for sufficiently large R; whence it follows that the integral along C_R'' approaches zero as R approaches infinity [since its modulus is less than $\pi R \times (1/\tfrac{1}{2}\,R^4)$ $= 2\pi/R^3$]. Therefore the residue method can be applied to evaluate the integral.

We next note that as long as $R > 2$ there are simple poles at i and $2i$ within the contour C_R considered, as well as two other poles at $-i$ and $-2i$ which do not concern us. The residues present no difficulty, as the relevant factor may be removed at once and the remaining expression evaluated, as in § 12.5(a), and we easily obtain

$I = $ real part of

$$2\pi i\{e^{i\cdot i}/(-1+4)(2i)+e^{i\cdot 2i}/(4i)(-4+1)\}$$
$$= \pi[e^{-1}/(+3)+e^{-2}/(-6)] \quad \text{or} \quad \pi(e^{-1}/3-e^{-2}/6).$$

Note that the imaginary part of the expression in curly brackets is zero. By the above, this implies that

$$\int_{-\infty}^{\infty} \frac{\sin mx}{(x^2+4)(x^2+1)}\, dx = 0.$$

Would it have been possible to deduce this easily by other means?

13.5. Jordan's Lemma

Consider now

$$\int_{-\infty}^{\infty} \frac{x \sin mx}{1+x^2}\, dx.$$

While it is true that $z/(1+z^2)$ approaches zero as z approaches infinity, and therefore as above that $e^{iz}z/(1+z^2)$ approaches zero as z approaches infinity in the upper half-plane, we still do not know that

$$\int_{C_R''} \frac{e^{iz}z}{1+z^2} dz$$

approaches zero as R approaches infinity, since in the estimates we have to include a factor πR to account for the length of C_R''. However, the following theorem can help us in such situations. We shall need a preliminary result.

PROPOSITION. *For $0 \leqslant \theta \leqslant \pi/2$ we have $2\theta/\pi \leqslant \sin \theta \leqslant \theta$.*

Proof. Let $y = \sin \theta/\theta$; then $dy/d\theta = (\theta \cos \theta - \sin \theta)/\theta^2$

$$= \cos \theta(\theta - \tan \theta)/\theta^2.$$

Now in the range $0 < \theta < \pi/2$ it is well known that $\tan \theta > \theta$ (see Exercise 13.5.1), so that y is a decreasing function, decreasing in fact from 1 to $2/\pi$. Therefore $1 \geqslant \sin \theta/\theta \geqslant 2/\pi$. The result follows. We can now proceed to the main theorem.

THEOREM (Jordan's lemma). *Let f be meromorphic in the upper half-plane and let $f(z)$ approach zero uniformly for all θ with $0 \leqslant \theta \leqslant \pi$, as $|z|$ approaches infinity. (This means explicitly that $|f(z)| < \varepsilon$ for $|z| = R > R_0(\varepsilon)$, for all amp $z = \theta$ with $0 \leqslant \theta \leqslant \pi$, where R_0 depends only on ε and not on θ.) Then for all $m > 0$, $\int_{C_R} e^{imz}f(z)\, dz$ approaches* *zero as R approaches infinity, where C_R is the upper semicircle $|z|=R$, $0 \leqslant \mathrm{amp}\ z \leqslant \pi$.*

Proof. We have to show that given $\varepsilon > 0$ we can choose R_0 such that

$$\int_{C_R} e^{imz} f(z)\, dz < \varepsilon \quad \text{for} \quad R > R_0\ (0 \leqslant \theta \leqslant \pi).$$

On letting $z = Re^{i\theta}$ the integral reduces to

$$I = \int_0^\pi e^{im(R\cos\theta + iR\sin\theta)} f(Re^{i\theta}) Rie^{i\theta}\, d\theta$$

$$= Ri \int_0^\pi e^{imR\cos\theta} e^{-mR\sin\theta} f(Re^{i\theta}) e^{i\theta}\, d\theta.$$

Given $\varepsilon > 0$, we shall suppose that R_0 is chosen so large that $|f(z)| < \varepsilon m/\pi$ if $R > R_0$, and it then follows that

$$|I| < \frac{R\varepsilon m}{\pi} \int_0^\pi e^{-mR\sin\theta}\, d\theta,$$

the neglected terms each having unit modulus.

But $\sin(\pi - \theta) = \sin\theta$, from which it easily follows that

$$\int_0^\pi e^{-mR\sin\theta}\, d\theta = 2\int_0^{\pi/2} e^{-mR\sin\theta}\, d\theta < 2\int_0^{\pi/2} e^{-mR2\theta/\pi}\, d\theta,$$

from the proposition above, and this final integral is equal to

$$\frac{\pi}{mR} \left[-e^{-2mR\theta/\pi} \right]_0^{\pi/2} = (\pi/mR)(1 - e^{-mR}).$$

It follows that $|I| < (R\varepsilon m/\pi)(\pi/mR) = \varepsilon$. This proves the theorem.

13.6. Order of Magnitude of a Function

A useful concept for examining quickly whether the value of a certain function is or is not negligible as z approaches infinity is that of *order of magnitude*. We say that $f(z)$ is of *order* R^m (as z approaches infinity) if there exist real constants K and R_0 such that $|f(z)| < KR^m$ whenever $|z| = R > R_0$. We write in this case $f(z) = 0 (R^m)$. Clearly a sufficient condition for $f(z)$ to approach zero as z approaches infinity is $f(z) = 0(R^m)$, where $m < 0$.

As examples, $\frac{1}{2}z+1$ and $3z-1/z$ are each $0(R^1)$ or $0(R)$, while $1/(1+z^2)$ is $0(R^{-2})$. (Give suitable values for K and R_0 in these examples.)

This notation enables us to deal speedily with the problem of the vanishing of an integral round an infinite semicircle C_R''. For example

$$(z^2+a^2)/(z^3+a^3) = 0(R^2)/0(R^3) = 0(R^{-1}),$$

and since $\pi R = 0(R)$ the product of the expressions will be of order R^0. All this tells us is that the integral round C_R'' has modulus less than some constant K if $R > R_0$. Therefore we see that the initial expression is not of a form suitable for integrating along the upper semicircle. On the other hand, $(z+a)/(z+b)(z^2+c^2)$ presents no difficulties, since it is of order R^{-2}, and $0(R)^{-2} \times 0(R) = 0(R^{-1})$, which implies that the integral will tend to zero as R tends to infinity.

13.7. Integration along Rectangular Paths

There is no special reason for selecting a semicircular path rather than a square path, say, in the evaluation of real infinite integrals by means of residues, except that the modulus of z remains constant along a semicircle and is therefore easier to handle. Useful results may often be obtained by considering, among other possibilities, rectangular paths with vertices at $-R, +R, +R+ia, -R+ia$, where R rends to infinity and a is some suitable constant. As an example let us consider the integration of f given by $f(z) = \operatorname{sech} z$ around the perimeter C of the rectangle with vertices $-R$, $+R$, $+R+i\pi$, $-R+i\pi$. This evidently reduces to four separate integrals, in two of which x is the only variable, and in the other two of which y is the only variable (where $z = x+iy$). In fact we have

$$\int_C \operatorname{sech} z \, dz = \int_{-R}^{R} \operatorname{sech} x \, dx + \int_{0}^{\pi} \operatorname{sech} (R+iy)i \, dy + \int_{+R}^{-R} \operatorname{sech} (x+i\pi) \, dx$$

$$+ \int_{\pi}^{0} \operatorname{sech} (-R+i\pi)i \, dy.$$

$$= I_1 + I_2 + I_3 + I_4, \text{ say.}$$

Examine first the integral

$$I_2 = \int_{0}^{\pi} \operatorname{sech} (R+iy) \, dy.$$

We have

$$|I_2| \leqslant |2/(e^R e^{iy} + e^{-R} e^{-iy})|_{\max} \pi \quad \text{(by § 9.5)}.$$

Now

$$|e^R e^{iy} + e^{-R} e^{-iy}| \geqslant |e^R e^{iy}| - |e^{-R} e^{-iy}| \quad \text{(see § 1.14)}$$
$$= e^R - e^{-R},$$

since

$$|e^{iy}| = |e^{-iy}| = 1.$$

Hence

$$|I_2| \leqslant 2\pi/(e^R - e^{-R}),$$

which evidently approaches zero as R approaches infinity. Similarly, with I_4. So we are now left with

$$\int_{-\infty}^{+\infty} [\operatorname{sech} x - \operatorname{sech} (x+i\pi)] \, dx = 2\pi i \sum \mathcal{R}.$$

In order to find $\sum \mathcal{R}$, the sum of the residues, we have first to find the poles of f. These are the points z for which $e^z + e^{-z} = 0$ or $e^{2z} = -1 = e^{(2n+1)\pi i}$. (Check that these are poles and not essential singularities.) (See § 12.3.)

Thus the only pole within C is given by $z = i\pi/2$, and the residue at this (simple) pole is easily seen, using l'Hôpital's rule, to be $1/\sinh{(i\pi/2)}$. This further reduces to $1/i \sin{(\pi/2)}$, which gives finally $\mathcal{R} = -i$. Therefore the integral itself is found to be of value $2\pi i(-i) = 2\pi$.

However, returning to the integral, we notice that

$$\cosh{(x+i\pi)} = \cosh x \cosh i\pi + \sinh x \sinh i\pi$$
$$= \cosh x \cos \pi + i \sinh x \sin \pi = -\cosh x,$$

so that our final calculation can be reduced to

$$2 \int_{-\infty}^{\infty} \operatorname{sech} x \, dx = 2\pi \quad \text{or} \quad \int_{-\infty}^{\infty} \operatorname{sech} x \, dx = \pi.$$

While this particular result can easily be derived by elementary methods, it will be found from the exercises that this method of residues can be used to obtain results not easily obtainable otherwise.

13.8. Rational Functions

As we have seen, a rational function $f(z) = p(z)/q(z)$ (where we assume p and q to have no common factor) has only a finite number of poles in the complex plane [corresponding to the zeros of $q(z)$] and has a pole of order m at infinity if the degree of $q(z)$ is n and that of $p(z)$ is $n+m$ ($m, n > 0$). These are the only singularities of f.

The following theorem shows that the converse of this statement is true. It illustrates the important part that singularities play in complex variable theory.

THEOREM. *If a function has no singularities other than poles either in the complex plane or at infinity, then it is a rational function.*

Proof. Let f be the function under discussion. Since there is at most a pole at infinity, the principal part at infinity will be of the form

$$b_1 z + b_2 z^2 + \ldots + b_n z^n$$

(where any or all of the b_r may be zero). Consider a circle C with centre the origin and radius R, the radius being large enough to include all finite poles within it; such a circle exists since the pole at infinity—if there is one—has to be isolated. If there is no pole at infinity, then $f(1/z)$ is regular at $z = 0$ and hence in a neighbourhood of the origin—thus there is a circle outside which f is regular. From the Bolzano–Weierstrass property, and the fact that poles are isolated, it follows that there will be only a finite number of poles within C (cf. § 2.4). Let these be $z_1, z_2, \ldots, z_p$. At each of these the expansion of the function will have a principal part; we will denote the principal part at z_r by P_r. We now consider the function F defined by

$$F(z) = f(z) - \sum_{r=1}^{p} P_r - b_1 z - b_2 z^2 - \ldots b_n z^n.$$

This function F is evidently regular in the entire plane, and also at infinity, since wherever we expand F in the form of a Laurent series, the principal part will vanish in view of the terms we have introduced. Thus $F(1/z)$ is regular (and hence continuous) at 0, so there exists a circle $C'(|z| = r$, say) and a constant M_1 such that $|F(1/z)| < M_1$ for z on or inside C'. Hence there is a circle $C''(|z| = 1/r)$ such that $|F(z)| < M_1$ for z on or outside C''. However, since F is regular (and hence continuous) everywhere, there is a constant M_2 such that $|F(z)| < M_2$ for z on or inside C''. (See Exercise 13.8.2.)

Thus $|F(z)| \leqslant \max(M_1, M_2)$ everywhere, and so by Liouville's theorem (Chapter 10) $F(z)$ is a constant—say $F(z) = K$.
It follows that

$$f(z) = K + \sum_{1}^{p} P_r + \sum_{1}^{n} b_r z^r.$$

Now the terms may be brought to a common denominator, and the whole expression reduces to the form $p(z)/q(z)$, where $p(z)$ and $q(z)$ are polynomials. This proves the theorem.

Exercises

In the following examples, all integrals along circles or circular arcs are to be taken in a counter-clockwise sense.

13.1.1. Let $f(z) = 1/z^2 + 1/(z^2+1)$ and C be the circle $|z| = 1.5$. Identify the poles of f, determine suitable circles C_1, etc., as in § 13.1, and carry out the integration of f around these circles from first principles. Hence show that $\int_C f(z) \, dz = 0$.

Check by applying Cauchy's residue theorem.

13.1.2. Give the Laurent series expansion of

$$f(z) = 1/(z^2+a^2) + 1/(z+b)^2 \quad (z \neq 0, b \neq \pm ia)$$

about each of the points ia, $-ia$, $-b$. State whether the poles are simple, double etc. What is the situation if $b = ia$?

13.1.3. Evaluate the following integrals along the contours stated:

(a)
$$\int_C \frac{1}{z^2+4} \, dz,$$

where C is (i) the circle $|z| = 1$, (ii) the circle $|z| = 4$, (iii) the upper semicircle of $|z| = 3$, followed by the segment of the real axis from $x = -3$ to $x = +3$.

(b)
$$\int_C \frac{1}{z^2-9} \, dz,$$

where C is the upper semicircle $|z| = 1$ followed by the lower semicircle $|z-\tfrac{1}{2}| = 1\tfrac{1}{2}$, and the segment of the real axis from $x = 2$ to $x = 1$. Hence deduce the integral along the curved section of this contour.

(c)
$$\int_C \frac{z+i}{\sin z} \, dz,$$

where C is the circle $|z| = 1$.

(d)
$$\int_C \coth z \, dz,$$

where C is (i) the circle $|z| = 1$, (ii) the circle $|z+2| = 1$, (iii) the circle $|z-1| = 2$.

13.2.1. Evaluate the integrals with respect to θ, between the limits $\theta = 0$, $\theta = 2\pi$, of the following functions F given by $F(\theta) =$

(a) $1/(1+a \sin \theta)$, $-1 < a < 1$;

(b) $\cos 2n\theta/(1+2a \cos \theta + a^2)$, $-1 < a < 1$, $n = 0, 1, 2, \ldots$;

(c) $\cos^2 \theta/(1+2a \cos \theta + a^2)$;

(d) $a/(a^2 + \sin^2 \theta)$;

(e) $\cos^6 \theta$.

Experiment with some other simple forms, e.g. $1/(a^2+b^2\cos^2\theta)$, $\sin m\theta$, $\tan m\theta$, noting any limitations on your constants.

13.3.1. Evaluate the following integrals (p and q are real positive constants):

(a) $\displaystyle\int_{-\infty}^{+\infty} dx/(x^4+16)$;

(f) $\displaystyle\int_{-\infty}^{+\infty} \frac{x^7}{(x^4+64)^2}\,dx$;

(b) $\displaystyle\int_{-\infty}^{+\infty} dx/(x^4+p^4)$;

(g) $\displaystyle\int_{0}^{\infty} dx/(x^4-x^2+1)$;

(c) $\displaystyle\int_{-\infty}^{+\infty} dx/(x^2+1)^2$;

(h) $\displaystyle\int_{-\infty}^{+\infty} \frac{x^3}{(x^4+x^2+1)^2}\,dx$;

(d) $\displaystyle\int_{0}^{\infty} dx/(x^2+25)^2$;

(i) $\displaystyle\int_{-\infty}^{+\infty} dx/(x^2+4)(x^2-2x+2)$.

(e) $\displaystyle\int_{0}^{\infty} dx/(x^2+q^2)^3$;

13.3.2. In none of the above examples do any of the poles of the associated complex function $f(z)$ fall on the real line. State what would happen to $f(x)$ at a pole x_0 on the real line. Could the integral be evaluated by this method in such a case? Examine $f(x) = 1/(x-1)^2$, $1/(x^2+1)(x-2)$ for example. (Note: It might be possible to exclude the pole on the real axis from a closed contour by means of a small semicircle indenting the upper semicircle.)

13.3.3. What difficulty arises in attempting to employ the given methods of contour integration to evaluate

$$\int_{-\infty}^{\infty} \frac{x^3}{x^4+x^2+1}\,dx.$$

13.3.4. For each of the following functions find the least upper bound of $|f(z)|$ on the circle $|z| = R$ (for sufficiently large R, say $R > 2$); state the value or values of amp z which will cause $|f(z)|$ to assume the value of the l.u.b.:

(a) $1/z^3$;

(d) $(z-1/z)$;

(b) $1/(z^2+4)$;

(e) $1/(z^4-z^2+1)$.

(c) $z^3/(z^3+16)$;

13.4.1. Prove that for each θ ($0 < \theta < \pi$), each positive constant k, and each integer n the modulus of $\exp(Rk\sin\theta)/R^n\sin\theta$ approaches infinity as $|R|$ approaches infinity.

13.4.2. Evaluate the following integrals:

(a) $\displaystyle\int_{-\infty}^{\infty} \frac{\cos x}{x^2+4}\,dx;$

(e) $\displaystyle\int_{-\infty}^{\infty} \frac{dx}{(x^2+a^2)\sec x}\ ;$

(b) $\displaystyle\int_{-\infty}^{\infty} \frac{x\sin x}{(x^2+1)^2}\,dx;$

(f) $\displaystyle\int_{-\infty}^{\infty} \frac{\cos mx}{(x^2+a^2)(x^2+b^2)}\,dx\ (0 < a < b);$

(c) $\displaystyle\int_{-\infty}^{\infty} \frac{\cos^2 x}{(x^2+1)^2}\,dx;$

(g) $\displaystyle\int_{-\infty}^{\infty} \frac{\cos mx}{(x^2+a^2)^2}\,dx.$

(d) $\displaystyle\int_{-\infty}^{\infty} \frac{\cos mx}{(x^2+1)^2}\,dx;$

13.4.3. Experiment with further integrations of the above types; consider carefully in each case the behaviour of $f(z)$ as $|z|$ approaches infinity.

13.5.1. Show by means of a diagram that $\theta < \tan\theta$ for $0 < \theta < \pi/2$. What happens when $\theta > \pi/2$?

13.5.2. Prove, by dividing the integral into two separate integrals, that if $f(\pi-\theta)=f(\theta)$, then

$$\int_0^{\pi} f(\theta)\,d\theta = 2\int_0^{\pi/2} f(\theta)\,d\theta.$$

13.5.3. Prove that if $f(x) < F(x)$ whenever $a < x < b$, and f, F are continuous, then

$$\int_a^b f(x)\,dx < \int_a^b F(x)\,dx.$$

13.6.1. Give the order of magnitude of the following functions f in terms of $R = |z|$ as z approaches infinity; give also a suitable K and R_0 such that $|f(z)| < KR^n$ for $R > R_0$, where $0(R^n)$ is the order of magnitude:

$f(z) =$ (a) $z^2-5z-23;$

(b) $z/(z+16);$

(c) $(z^2+4)(z^2-4);$

(d) $(z+1)/(z^3-1);$

(e) $(z^2+3z+5)/(z^2-z-17);$

(f) $\sin(1/z)/(z+1);$

(g) $z\exp(1/z).$

13.6.2. Consider how to define the order of magnitude of a complex function f as (i) $z \to 0$, (ii) $z \to a$.

13.7.1. By integrating a suitable complex function around the perimeter of a rectangle with vertices $-R$, R, $R+2\pi$, $-R+2\pi$, prove that if $k < 1$ then

$$\int_{-\infty}^{\infty} \frac{e^{kx}}{e^x+1}\, dx = \pi \operatorname{cosec} k\pi.$$

13.7.2. By integrating $f(z) = z/(\frac{1}{2} - e^{-iz})$ around the rectangle with vertices $-\pi$, π, $\pi + in$, $-\pi + in$, deduce that

$$\int_{0}^{\pi} \frac{x \sin x}{5/4 - \cos x}\, dx = 2\pi \log (3/2).$$

13.7.3. By integrating $e^{kz}/\sinh \pi z$ around the rectangle with vertices $z = -R$, R, $R+i$, $-R+i$, indented suitably at 0 and i, derive a value for

$$\int_{0}^{\infty} \frac{\sinh kx}{\sinh \pi x}\, dx$$

in terms of $\tan k/2$. (Assume $|k| < \pi$).

13.8.1. If f in § 13.8 is given by

$$f(z) = \frac{1 + 2z + 3z^2}{(2+z)(1-2z)^2},$$

determine the positions and nature of the poles, considering also the character of the point at infinity. Determine also P_1, P_2 and F for this function (as in § 13.8), showing that $F(z)$ reduces to a constant.

13.8.2. Prove, by considering the interior and exterior of a sufficiently large circle, that a function which is continuous in the whole finite plane and bounded at infinity is bounded everywhere. (Compare this result with the maximum–modulus principle in the following chapter.)

(Note: a function f is said to be *bounded* over a given domain D if there exists some R such that $|f(z)| < R$ for all z in D. The function is said to be *bounded at infinity* if there exist R, K such that $|f(z)| < R$ for all z such that $|z| > K$.)

CHAPTER 14

MISCELLANEOUS THEOREMS

WE INCLUDE in this chapter certain results, interesting in themselves, which were not needed in the development of the previous chapters.

14.1. The Maximum–Modulus Principle

This states that if f is regular within and on a closed contour C, then (unless $f(z)$ is constant) $|f(z)|$ attains its maximum value on, and not within, C.

The proof we shall give requires a lemma.

LEMMA. *If ψ is a continuous function of a real variable x over the interval $[a, b]$, and if $\psi(x) \leqslant k$, while $\{1/(b-a)\} \int_a^b \psi(x)\,dx \geqslant k$, then $\psi(x) = k$ precisely.*

Proof of lemma. Assume that $\psi(x_1) < k$ for some point x_1. Say $\psi(x_1) = k - c$ ($c > 0$). Then, since ψ is continuous, there will be an interval $[x_1 - \delta, x_1 + \delta]$ over which $\psi(x) < k - \frac{1}{2}c$.

Breaking up the integral into three stages,

$$\int_a^{(x_1-\delta)} + \int_{(x_1-\delta)}^{(x_1+\delta)} + \int_{(x_1+\delta)}^{b} ,$$

we easily see that

$$\int_a^b \psi(x)\,dx < (x_1 - \delta - a)k + 2\delta(k - \tfrac{1}{2}c) + (b - x_1 - \delta)k$$
$$= (b-a)k - \delta c \quad \text{where} \quad \delta > 0,\, c > 0,$$
$$< (b-a)k.$$

But this gives $\{1/(b-a)\} \int_a^b \psi(x)\,dx < k$, contrary to our assumption.

225

Thus $\psi(x)$ can be nowhere less than k. Since by assumption $\psi(x)$ is nowhere greater than k, we have $\psi(x) = k$ exactly. This proves the lemma.

We now state the main theorem.

THEOREM. (Maximum–modulus principle.) *Let C be a simple closed curve in the complex plane, and let D denote C together with all the point s inside C. Then, unless $f(z)$ is constant, the maximum value of $|f(z)|$ for z belonging to D will be attained on C and not within the interior of D.*

Proof. Let us assume the theorem to be untrue, so that $|f(z)|$ attains its maximum at an interior point z_0 of D. We take a small circle C_1 lying entirely within D, with centre z_0 and radius r. We now write $z - z_0 = re^{i\theta}$ (so that z lies on C_1), $f(z) = f(z_0)\,\varrho e^{i\phi}$. Here ϱ and ϕ are real and both are functions of θ.

Now from Cauchy's integral formula (§ 10.4),

$$f(z_0) = \frac{1}{2\pi i} \int_{C_1} \frac{f(z)\,dz}{z - z_0}$$

$$= \frac{1}{2\pi i} \int_0^{2\pi} \frac{f(z_0)\varrho e^{i\phi}\, ire^{i\theta}\, d\theta}{re^{i\theta}}$$

$$= \frac{f(z_0)}{2\pi} \int_0^{2\pi} \varrho e^{i\phi}\, d\theta,$$

whence

$$1 = \frac{1}{2\pi} \int_0^{2\pi} \varrho e^{i\phi}\, d\theta. \tag{1}$$

Replacing $e^{i\phi}$ by its modulus in eqn. (1) we deduce that

$$1 \leqslant \frac{1}{2\pi} \int_0^{2\pi} \varrho\, d\theta. \tag{2}$$

But if $|f(z_0)|$ is a maximum, it follows that $\varrho \le 1$. From this fact, and eqn. (2), application of the above lemma shows that $\varrho = 1$ precisely for all values of θ. Returning to eqn. (1) and equating real parts now gives

$$1 = \frac{1}{2\pi} \int_0^{2\pi} \cos \phi \, d\theta.$$

But $\cos \phi \le 1$, so again applying the lemma gives $\cos \phi = 1$, $\phi = 0$ (or $2n\pi$). These values of ϱ and ϕ show that, on $C_1, f(z) = f(z_0)$. But r was arbitrary so long as C_1 lay in D. Hence $f(z)$ is constant in a neighbourhood of z_0.

We now show that it follows from this that $f(z)$ is constant everywhere in the interior of D, and hence (by continuity) everywhere in D. Let K denote the set of all points z in the interior of D such that $f(z) = f(z_0)$. We shall show that K is in fact the whole of the interior of D by supposing the contrary and deducing a contradiction.

If K is not the whole of the interior of D we can find a point z_1 inside C and not an element of K. Let g be a path joining z_0 to z_1 and lying entirely inside C. It is easy to show (see Exercise 2.8.6) that there is a point z_2 on g such that z_2 is a boundary point of K. If z_2 is an element of K [i.e. if $f(z_2) = f(z_0)$] then by replacing z_0 by z_2 in all the above we see that z_2 has a neighbourhood of points z with $f(z) = f(z_2)$, i.e. a neighbourhood consisting of points which are elements of K, which contradicts the fact that z_2 is a boundary point of K. On the other hand, if z_2 is not an element of K, we have $f(z_2) \ne f(z_0)$. It then follows from the continuity of f that z_2 must have a neighbourhood of points z for which $f(z) \ne f(z_2)$, i.e. a neighbourhood of points not elements of K, which again contradicts the fact that z_2 is a boundary point of K. This contradiction completes the proof.

This proof is based on the first of three proofs of the maximum-modulus principle given in reference 7. Others may be found in references 2 and 5.

14.2. A Theorem concerning Poles and Zeros
of Meromorphic Functions

THEOREM. *Let f be regular and with no zero at any point on a contour C and suppose that f is meromorphic within C. Then if N is the number of zeros of f within C and P the number of poles within C, each zero or pole counted the appropriate number of times according to its multiplicity, we have*

$$\frac{1}{2\pi i} \int_C \frac{f'(z)\,dz}{f(z)} = N - P.$$

Proof. Let a be a zero of order m; then we may write

$$f(z) = (z-a)^m \,\phi(z),$$

where ϕ is regular and not zero on a neighbourhood of a. Taking logarithms and differentiating, or direct verification, gives

$$f'(z)/f(z) = m/(z-a) + \phi'(z)/\phi(z).$$

The last term is regular near a, and so $f'(z)/f(z)$ has a simple pole at a (with residue m).

In exactly the same way, if b is a pole of order n of f we may write for points in a neighbourhood of b,

$$f(z) = (z-b)^{-n} \, F(z),$$

where F is regular and non-zero in this neighbourhood. We hence see that f'/f has a simple pole, with residue $-n$, at b. Since f'/f is regular at all other points within or on C, a simple application of the residue theorem at once gives

$$\int_C \frac{f'(z)}{f(z)}\,dx = 2\pi i [\sum m + \sum (-n)],$$

the summations being taken over all zeros and poles within C. The result follows.

Since at points z where f is regular we have

$$\frac{d}{dz}[\log f(z)] = f'(z)/f(z),$$

it follows that the value of the integral

$$\int_C \frac{f'(z)}{f(z)}\,dx$$

represents the total variation in the value of $\log f(z)$ as z traverses C once in the counter-clockwise direction (see § 9.4). Now $\log f(z) = \log |f(z)| + i$ amp $f(z)$, and the modulus of $f(z)$ is unaltered as z goes once round C. Thus only the amplitude is (possibly) altered, and hence for a regular function f we may write our previous result in the form

$$2\pi i N = \Delta_C i \text{ amp} f(z),$$

where N is the number of zeros of f (there are no poles) within C, and Δ_C denotes the variation round C of the regular function being considered.

Thus

$$N = \frac{1}{2\pi}\Delta_C \text{ amp } f(z).$$

Evidently this will always reduce to an integer, since the variation in the amplitude must be a multiple of 2π.

14.3. Rouché's Theorem

This theorem provides a most attractive way of demonstrating that every polynomial of degree n has exactly n zeros in the complex plane.

THEOREM. *Let f and g be two functions of z which are regular within and on a simple closed contour C. Suppose $|g(z)| < |f(z)|$ for all z on C. Then f and $f+g$ have the same number of zeros within C.*

Proof. We note first that neither f nor $f+g$ has a zero on C, since $|f(z)| > |g(z)| \geqslant 0$ on the contour, and

$$|f(z)+g(z)| \geqslant |f(z)| - |g(z)| > 0.$$

Let N and N_1 be the number of zeros of f, and of $f+g$, respectively, within C. Then, from the previous theorem,

$$2\pi N = \Delta_C \operatorname{amp} f(z),$$
$$2\pi N_1 = \Delta_C \operatorname{amp} [f(z)+g(z)]$$
$$= \Delta_C \operatorname{amp} \{f(z)[1+g(z)/f(z)]\}$$
$$= \Delta_C \{\operatorname{amp} f(z)+\operatorname{amp} [1+g(z)/f(z)]\}$$

since the amplitude of a product is the sum of the amplitudes of the actors)

$$= \Delta_C \operatorname{amp} f(z)+\Delta_C \operatorname{amp} [1+g(z)/f(z)].$$

Now since $|g(z)/f(z)| < 1$ by hypothesis, the point $1+g(z)/f(z)$ lies within the circle with centre 1 and radius 1, namely $|w-1| = 1$, and so (as is seen readily from Fig. 14.1) amp $[1+g(z)/f(z)]$ lies between $-\frac{1}{2}\pi$ and $+\frac{1}{2}\pi$. Therefore when z describes C, amp $[1+g(z)/f(z)]$ cannot

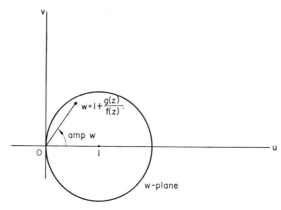

Fig. 14.1. If w lies within $|z-1| = 0$, then amp w lies between $\pi/2$ and $-\pi/2$,

vary by a non-zero multiple of 2π, and thus it returns to its original value, i.e. Δ_C amp $[1+g(z)/f(z)] = 0$. It follows at once that $2\pi N = 2\pi N_1$, whence $N = N_1$, and the theorem is proved.

14.4. "The Fundamental Theorem of Algebra"

THEOREM. *Every non-constant polynomial with real or complex coefficients has at least one zero.*

Instead of proving this theorem we shall, by the use of Rouché's theorem, obtain a stronger result.

THEOREM. *Every polynomial of degree n with real or complex coefficients has n zeros.*

Proof. Let the polynomial be

$$a_0 z^n + a_1 z^{n-1} + \ldots + a_{n-1} z + a_n.$$

Let

$$f(z) = a_0 z^n \quad \text{and} \quad g(z) = a_1 z^{n-1} + \ldots + a_{n-1} z + a_n.$$

Let z lie on a circle C given by $|z| = R$, where R is taken so large that $|f(z)| > |g(z)|$ on C. This may always be done, since

$$|g(z)| \leqslant |a_1|R^{n-1} + \ldots + |a_n|$$
$$< R^{n-1}(|a_1| + \ldots + |a_n|) \quad \text{for} \quad R > 1$$
$$< |a_0 z^n|$$

provided that

$$R > (|a_1| + |a_2| + \ldots + |a_n|)/|a_0|.$$

(Verify this.)

Now f and g are certainly regular inside and on C, so that they fulfil the conditions of Rouché's theorem. It follows that since f has precisely n zeros inside C (an n-fold zero at the origin), so does $f+g$. As $f+g$ has no zeros outside C—from the inequality above, the modulus of the first term exceeds that of all the other terms together for z on and outside C—the theorem follows.

APPENDIX A

MULTIPLICATION OF COMPLEX NUMBERS

WHEN trying to find a good definition for multiplication of p-numbers (§ 1.8) we had two conditions which we wished to be satisfied:

(1) if z_1 happens to be real, then $z_1 \times z_2$ should be as already defined (in § 1.6),

(2) $|z_1 \times z_2|$ should be equal to $|z_1| \times |z_2|$.

Let us now see what would be implied by omitting (2), and replacing it by three other very reasonable conditions:

(2′) $|z_1 \times z_2|$ should be less than or equal to $|z_1| \times |z_2|$ [this is a direct analogue of the triangle inequality (§ 1.14) which we know to be true for addition];

(2″) multiplication should be commutative, i.e. $z_1 \times z_2 = z_2 \times z_1$ for all z_1, z_2;

(2‴) multiplication should be distributive over addition, i.e.

$$\left. \begin{array}{l} z_1 \times (z_2 + z_3) = z_1 \times z_2 + z_1 \times z_3 \\ (z_2 + z_3) \times z_1 = z_2 \times z_1 + z_3 \times z_1 \end{array} \right\} \text{ for all } z_1, z_2, z_3 \text{ (see § 1.11.).}$$

Let $z_1 = (a, b)$, $z_2 = (c, d)$ be two arbitrary p-numbers. We have

$$(a, b) \times (c, d) = [a(1, 0) + b(0, 1)] \times [c(1, 0) + d(0, 1)].$$

If we now expand by means of the distributive laws (2‴), at the same time writing $a(1, 0)$ as a and $c(1, 0)$ as c and using the commutative

232

law (2″), we obtain

$$(a, b) \times (c, d) = ac + (bc + ad)(0, 1) + bd(0, 1) \times (0, 1). \tag{1}$$

Let us denote the p-number $(0, 1) \times (0, 1)$ by (x, y), where x and y are real numbers yet to be determined. The right-hand side of eqn. (1) may then be written

$$ac + (bc + ad)(0, 1) + bd[x(1, 0) + y(0, 1)]$$

or as the p-number $(ac + bdx,\ bc + ad + bdy)$.

Now by our assumption (2′) we have

$$|(a, b)| \times |(c, d)| \geqslant |(a, b) \times (c, d)|$$

and this gives, on squaring each side,

$$(a^2 + b^2)(c^2 + d^2) \geqslant (ac + bdx)^2 + (bc + ad + bdy)^2$$

which easily reduces to

$$b^2 d^2 \geqslant 2abcd(1 + x) + b^2 d^2 x^2 + 2(bc + ad)bdy + b^2 d^2 y^2. \tag{2}$$

Since we require this inequality to hold for all real a, b, c, d, it must certainly hold for the special case in which $b = d = 1$, which we shall now examine.

Substitution of these values reduces the inequality (2) to

$$1 \geqslant 2ac(1 + x) + x^2 + 2(c + a)y + y^2$$

or

$$1 - x^2 - y^2 \geqslant 2ac(1 + x) + 2(c + a)y. \tag{3}$$

This is to hold for all a and c. Consider first the possibility that $1 + x$ is positive. Then, if we let $a = c = N$, the right-hand side of inequality (3) approaches infinity with N, whatever value y holds, whereas the left-hand side is not greater than 1. On the other hand, if $1 + x$ is negative, and we let $a = -c = N$, we obtain on the right-hand side the single term $-2N^2(1 + x)$, which again increases towards $+ \infty$ with N. Hence the inequality (3) is not possible unless $x = -1$.

Since the right-hand side then reduces to $2(c+a)y$, and c and a may be any real numbers whatever, the inequality can only be made to hold generally if $y = 0$. (It then, in fact, reduces to the equality $0 = 0$.) Thus we have proved $x = -1$, $y = 0$ to be the only possible values to consider.

Using these values we now have

$$(a, b) \times (c, d) = (ac + bdx, bc + ad + bdy) = (ac - bd, bc + ad),$$

which is the formula we obtained in § 1.8.

We have therefore shown that the definition of multiplication which we assumed in § 1.8, bearing in mind conditions (1) and (2), is in fact the only definition which satisfies the quite reasonable alternative conditions (1), (2'), (2''), (2''').

[See J. I. Nassar, On multiplication of ordered pairs of real numbers, *Mathematical Gazette* **50** (372) 118 (1966).]

APPENDIX B

GROUPS

WHEN considering rules for addition and multiplication of complex numbers, we observed that these operations had certain properties in common (associativity, existence of inverses, etc.). These properties in fact occur in many different contexts far removed from real or complex numbers, and are the basic properties of the abstract mathematical concept known as a *group*.

DEFINITION. *A group is a set G together with an operation $*$ which associates with each ordered pair x, y of elements of G an element x_*y in G, and has the following properties:*

(a) *it is associative, i.e.* $x_*(y_*z) = (x_*y)_*z$ *for all elements x, y, z in G;*

(b) *there is a unique element e (the "identity element") in G such that $e_*x = x_*e = x$ for all elements x in G;*

(c) *to each element x in G there corresponds an element, denoted by x^{-1}, such that $x_*x^{-1} = x^{-1}_*x = e$. This element x^{-1} is called the "inverse" of x.*

[The inverse of an element is unique, since if $x_*x' = x'_*x = e$ we deduce that

$$x' = x'_*e = x'_*(x_*x^{-1}) = (x'_*x)_*x^{-1} = e_*x^{-1} = x^{-1}.]$$

It is possible for an element to be its own inverse, as, for example, in the case of e.

If, in addition to the properties (a), (b), and (c) we also have

(d) $x_*y = y_*x$ *for all x, y in G,*

then we say the group is an *abelian group* (or *commutative group*).

Exercises

Show that the following are groups:
(i) integers with addition;
(ii) non-zero rational numbers with multiplication;
(iii) real numbers with addition;
(iv) the symbol e with $e_*e = e$;
(v) the set of symbols $\{a, b, c\}$ with $a_*a = a$, $a_*b = b_*a = b$, $a_*c = c_*a = c$, $b_*b = c$, $b_*c = c_*b = a$, $c_*c = b$.

Show that the following are not groups, and state why not:
(i) positive integers with addition;
(ii) non-zero rational numbers with addition;
(iii) real numbers with multiplication;
(iv) the set of symbols $\{a, b\}$ with $a_*a = a_*b = b_*a = a$;
(v) the set of symbols $\{a, b, c\}$ with $a_*a = a$, $a_*b = b_*a = b$, $a_*c = c_*a = c$, $b_*b = c$, $b_*c = c_*b = a$, $c_*c = a$.

APPENDIX C

THE MILNE–THOMPSON METHOD

IN APPLICATIONS of complex variable theory it is often necessary to be able to reconstruct a regular function when its real part (or imaginary part) alone is given.

If f is regular and $f(x+iy) = u(x, y)+iv(x, y)$, then the function f_1 given by $f_1(x+iy) = u(x, y)+iv(x, y) +ic$ (where c is real) is also regular and has the same real part as f. Therefore if the real part of f is given we can at most hope to determine f to within an arbitrary imaginary constant. Similarly, if the imaginary part v is given we can determine f only to within a real constant. The following method is due to Milne–Thompson.

We shall suppose for simplicity that f is regular in the whole plane. When f is regular in some smaller domain D, the method is the same, but the justification needs to be worded with additional care.

Let
$$w(x, y) = f(x+iy) = u(x, y)+iv(x, y).$$
Then, as we have seen (§ 4.4),
$$df(z)/dz = u_x(x, y)+iv_x(x, y)$$
$$= u_x(x, y)-iu_y(x, y)$$
by the Cauchy–Riemann equations.

Suppose that $u_x(x, y)$ and $u_y(x, y)$ are expressible as formulae in which it is meaningful to substitute z instead of x [e.g. x^2, $e^x \cos y$, $(x^2-y^2)/(x^2+y^2)^2$], and let $u_x(z, 0)$ and $u_y(z, 0)$ denote the results of substituting z for x and 0 for y in u_x and u_y. Suppose also that $u_x(z, 0)$ and $u_y(z, 0)$ are regular functions of z. Then
$$df/dz \quad \text{and} \quad u_x(z, 0)-iu_y(z, 0)$$

are both regular functions of z, and they agree for all real values of z. However, it can be proved by considering power series expansions that two regular functions which agree for all values of z lying on the real axis must be identical. Therefore we have

$$df/dz = u_x(z, 0) - u_y(z, 0)$$

for all z, and so $f(z)$ can be determined by integration.

For an example of this method, suppose

$$u(x, y) = y/(x^2 + y^2).$$

Then

$$u_x = -2xy/(x^2 + y^2)^2 \quad \text{and} \quad u_y = (x^2 - y^2)/(x^2 + y^2)^2.$$

When we replace x by z and y by 0, the expression for u_x reduces to 0, and that for u_y to $1/z^2$.

Hence the required integration is given by

$$f(z) = \int (0 - i/z^2)\, dz = i/z + c.$$

Thus we are led to the result

$$f(z) = i/z + c = iz/z\bar{z} + c = i(x - iy)/(x^2 + y^2) + c$$

or

$$w = y/(x^2 + y^2) + ix/(x^2 + y^2) + c.$$

This evidently gives the correct u if we take the arbitrary constant as imaginary, and also the corresponding v.

MISCELLANEOUS EXERCISES

Section A. Complex Numbers. Convergence

1. Prove that

$$(\cos\theta + i\sin\theta)^n = \cos n\theta + i\sin n\theta \quad (n = \pm1, \pm2, \pm3, \ldots).$$

By first solving the equation $\cos 3\theta + \sin 3\theta = 0$, or otherwise, show that the roots of the equation $t^2 + 4t + 1 = 0$ are

$$t = -\tan\frac{\pi}{12} \quad \text{and} \quad t = -\tan\frac{5\pi}{12}.$$

2. Explain how complex numbers are represented by points in the Argand diagram, and explain how to find the points representing $z_1 + z_2$ and $z_1 z_2$ when the points representing z_1 and z_2 are given.

What conditions have to be satisfied by the complex number z in order that the points representing all integral powers of z should (i) lie on a circle with centre at the origin, (ii) be finite in number? Mark on the diagram the points which represent a number z such that there are only three distinct points in the sequence given by $z, z^2, \ldots$

3. Show that multiplication of a complex number $z = x + iy$ by i increases its argument by $\pi/2$ without altering its modulus. Illustrate this result geometrically in an Argand diagram.

If the points P and Q in an Argand diagram represent the complex numbers z_1 and z_2 respectively and O is the origin, show that if the triangle OPQ is isosceles and right angled at O, then $z_1^2 + z_2^2 = 0$.

Find the complex numbers represented by the vertices of a square if one vertex represents $3 + 3i$ and the centre of the square represents $1 + 2i$.

4. (i) Prove that the points which represent the roots of the equation

$$(1 - z)^n = z^n$$

in the Argand diagram are collinear.

(ii) State De Moivre's theorem, and prove it for integral indices, positive or negative.

Express $x^9 + 1$ as a product of one linear and four quadratic real factors.

5. Find the six distinct values of the sixth roots of unity.

Prove that $(1 + i)^6 = -8i$ and find the six complex values of $(-8i)^{1/6}$.

Indicate in an Argand diagram the position of the points representing these six values and mark in the same diagram the points representing $(1+i)^n$ for $n = 0$, 1, 2, 3, 4.

6. Define what is meant by the statements:

(a) $\sum_{n=1}^{\infty} u_n$ is convergent;

(b) $\sum_{n=1}^{\infty} u_n$ is absolutely convergent.

Quote an example of a series which is convergent but not absolutely convergent. State, without proof, whether or not there exist absolutely convergent series which are not convergent.
Establish the convergence or divergence of the series

$$\frac{1}{1.2}+\frac{1}{2.3}-\frac{1}{3.4}+\frac{1}{4.5}+\frac{1}{5.6}-\frac{1}{6.7}++-\cdots$$

7. Find the loci in the Argand diagram corresponding to the equations

(i) $$\left|\frac{z-1}{z+1}\right| = 2;$$

(ii) $$\arg\left(\frac{z-i}{z+i}\right) = \pi/2,$$

(iii) $$|z|+|z-6| = 10.$$

8. Explain what is meant by saying that the series $\sum_{n=1}^{\infty} u_n$ is absolutely convergent. Give an example of a series which is convergent but not absolutely convergent. Show that, if θ is not a multiple of π,

$$\sum_{n=1}^{N} \cos^n \theta \cos n\theta = \cos^{N+1} \theta \sin N\theta \operatorname{cosec} \theta.$$

Prove that, for $0 < \theta < \pi$, the series $\sum_{n=1}^{\infty} \cos^n \theta \cos n\theta$ is absolutely convergent and has zero sum.

9. Explain what is meant by the modulus and argument (or amplitude) of a complex number z.
If z_1 and z_2 are two complex numbers, show how you would represent z_1, z_1+z_2, $\bar{z}_1$, $z_1 z_2$ in an Argand diagram. ($\bar{z}$ denotes the complex conjugate of z.)
Show that the equation of the circle with centre (a, b) and radius r can be written as

$$z\bar{z}-(a-ib)z-(a+ib)\bar{z}+a^2+b^2 = r^2.$$

Hence, or otherwise, prove that every circle through the points $(-1, 0)$, $(1, 0)$ has an equation of the form

$$z\bar{z} + ikz - ik\bar{z} - 1 = 0,$$

where k is real.

10. Find the solutions of the equation $z^5 = 1$ and indicate their positions in the Argand diagram.

If α and β are two distinct solutions of the given equation, prove that the greatest possible value of $|\alpha + \beta|$ is $2\cos(\pi/5)$, and find its smallest possible value.

11. The Taylor expansions of the functions $\cos 2x$ and x^{-2} in powers of $(x - \pi/4)$ are given by

$$\cos 2x = \sum_{n=0}^{\infty} a_n \left(x - \frac{\pi}{4}\right)^n; \qquad x^{-2} = \sum_{n=0}^{\infty} b_n \left(x - \frac{\pi}{4}\right)^n.$$

Find a_n and b_n and state the range of values of x for which each expansion is valid. Evaluate

$$\sum_{n=0}^{\infty} |a_n|,$$

and find the function whose Taylor expansion is

$$\sum_{n=0}^{\infty} |b_n| \left(x - \frac{\pi}{4}\right)^n.$$

12. State and prove de Moivre's theorem for a positive integral index. Find all the solutions of the equation

$$z^6 + 1 = i\sqrt{3}.$$

Show that three of them are solutions of

$$\sqrt{2}\, z^3 + i\sqrt{3} = -1,$$

and find the cubic equation satisfied by the other three.

13. The points representing the complex numbers z_1 and z_2 lie respectively in the first and third quadrants of the Argand plane. Indicate graphically $|z_1 - z_2|$, $\arg(z_1 - z_2)$ and $\arg(z_2 - z_1)$.

Find the locus of z when

$$\text{(a)} \quad \left|\frac{z-i}{z+i}\right| = 1; \qquad \text{(b)} \quad \arg\frac{z-i}{z+i} = \frac{\pi}{2}.$$

Also show that, if

$$w = \frac{z-i}{z+i},$$

hen the region $|w| < 1$ corresponds to the upper z half-plane.

14. $\sum a_n$ is a series of strictly positive terms and

$$\frac{a_n}{a_{n+1}} \to l \quad (n \to \infty).$$

Show that the series is convergent if $l > 1$ and divergent if $0 \leqslant l < 1$. Show also that $l = 1$ is consistent with convergence and with divergence.

Discuss the convergence, for real values of x, of the series

$$\sum_{n=0}^{\infty} \frac{3.5\ldots(2n+1)}{2.4\ldots 2n} (1-x^2)^n.$$

15. Find all values of x for which each of the following series of real terms is convergent:

(a)
$$x + \frac{x^4}{2} + \frac{x^9}{3} + \ldots + \frac{x^{n2}}{n} + \ldots;$$

(b)
$$\frac{2x}{1!} + \frac{3x^2}{2!} + \ldots + \frac{(n+1)x^n}{n!} + \ldots.$$

16. In each of the five following pairs of statements, (a) and (b), one statement is true and the other is false. Write down which is the false statement, and demonstrate its falsity by means of a counter-example.

(i) (a) Every non-empty bounded set of rational numbers has a least upper bound which is a rational number.

(b) Every non-empty bounded set of real numbers has a greatest lower bound which is a real number.

(ii) $\{s_n\}$ and $\{t_n\}$ are convergent sequences such that $\lim_{n \to \infty} s_n = S$ and $\lim_{n \to \infty} t_n = T$.

(a) $\lim_{n \to \infty} s_n t_n = ST$;

(b) $\lim_{n \to \infty} \frac{s_n}{t_n} = \frac{S}{T}$.

(iii) (a) If $\sum u_n$ is convergent, then $\lim_{n \to \infty} u_n = 0$;

(b) If $\lim_{n \to \infty} u_n = 0$, then $\sum u_n$ is convergent.

(iv) (a) If $\lim_{n \to \infty} \left| \frac{u_n}{u_{n+1}} \right| = l > 1$, then $\sum u_n$ is absolutely convergent;

(b) if $\lim_{n \to \infty} \left| \frac{u_n}{u_{n+1}} \right| = 1$, then $\sum u_n$ is not convergent.

(v) (a) If $\sum a_n$ and $\sum b_n$ are series of positive terms, and $\lim_{n \to \infty} \frac{a_n}{b_n} = L$, where L is finite and non-zero, then $\sum a_n$ and $\sum b_n$ are either both convergent or both divergent;

(b) if $\sum a_n$ and $\sum b_n$ are series of real terms, and $\lim\limits_{n\to\infty} \left|\dfrac{a_n}{b_n}\right| = L$, where L is finite and non-zero, then $\sum a_n$ and $\sum b_n$ are either both convergent or both divergent.

17. By considering infinite series, or otherwise, justify the statement that, for real x, $e^{ix} = \cos x + i \sin x$.

If n is a positive integer, prove that $(e^{ix})^n = e^{inx}$, and indicate why this does not follow immediately from the index law: $(a^m)^n = a^{mn}$.

Solve completely the equation $z^6 + 1 = 0$, giving each of your answers in the form $a + ib$, where a and b are real.

18. What is meant by saying that a series of real terms $\sum\limits_{n=1}^{\infty} a_n$ is "absolutely convergent"?

Prove d'Alembert's ratio test for the convergence of a series $\sum\limits_{n=1}^{\infty} u_n$ of positive terms.

Explain how d'Alembert's test can be used to investigate the convergence of a series containing both positive and negative terms. State one other test for the convergence of a series of real terms.

Determine the range of convergence of each of the following series:

(a) $\sum\limits_{n=1}^{\infty} (-1)^n n^2 x^n$; (b) $\sum\limits_{n=1}^{\infty} \dfrac{(-1)^n x^n}{2n+1}$.

19. Prove that an absolutely convergent series is convergent.

Discuss the convergence and absolute convergence of the following series:

(i) $\sum \dfrac{\cos (n\pi/4)}{n(n+1)}$;

(ii) $\sum \dfrac{(-1)^n}{\sqrt[n]{n}}$;

(iii) $\sum (-1)^n \sin \dfrac{\pi}{n}$.

20. Assuming that every non-empty set of real numbers that is bounded above has a least upper bound, prove that:

(a) every non-empty set of real numbers that is bounded below has a greatest lower bound;

(b) every bounded infinite set of real numbers has at least one limit point (accumulation point).

Give examples to show that the greatest lower bound in (a) and the accumulation point in (b), may—but need not—be members of the set concerned.

Give a formal definition of the limit l of a convergent sequence $\{s_n\}$ of real numbers, in terms of numbers l, N and e. If s_n is the sum to n terms of the series

$$1 + \tfrac{1}{2} + \tfrac{1}{4} + \ldots + \left(\tfrac{1}{2}\right)^r + \ldots,$$

determine a suitable N if $\varepsilon = 0.01$.

21. Show that the series $\sum \dfrac{z^n}{n!}$ is convergent for all complex numbers z. If

$E(z) = \displaystyle\sum_{n=0}^{\infty} \dfrac{z^n}{n!}$ prove that

(i) $E(z_1) E(z_2) = E(z_1 + z_2)$;

(ii) $E(0) = 1$ and $E(z) \neq 0$ for all z;

(iii) $\dfrac{d}{dz} E(z) = E(z)$.

(State carefully the theorems on power series which you use.)

22. (i) Sketch the loci in the z-plane represented by:

(a) $\left| \dfrac{z-1}{z+1} \right| = 2$;   (b) $\arg\left(\dfrac{z-1}{z+1}\right) = \dfrac{\pi}{3}$.

(ii) Show that if the four points which represent the complex numbers z_1, z_2, z_3, z_4 are concyclic, then

$$\dfrac{(z_1 - z_2)(z_3 - z_4)}{(z_1 - z_4)(z_3 - z_2)} \text{ is real.}$$

23. (a) Determine whether the following series converge or diverge:

(i) $1 - \dfrac{6}{7} + \dfrac{8}{10} - \dfrac{10}{13} + \dfrac{12}{16} - \dfrac{14}{19} + \ldots$;

(ii) $\dfrac{1}{3} + \dfrac{1}{5^2} + \dfrac{1}{3^3} + \dfrac{1}{5^4} + \dfrac{1}{3^5} + \dfrac{1}{5^6} + \ldots$.

(b) Given that $\displaystyle\sum_{n=1}^{\infty} a_n$ is a convergent series of positive terms, prove that $\displaystyle\sum_{n=1}^{\infty} \dfrac{a_n}{1+a_n}$

and $\displaystyle\sum_{n=1}^{\infty} a_n^2$ are also convergent.

(c) Discuss the convergence, for real values of x, of the series

(i) $\displaystyle\sum_{n=0}^{\infty} \dfrac{1}{n}\left(\dfrac{x}{2}\right)^{2n}$;   (ii) $\displaystyle\sum_{n=1}^{\infty} \left(x^n + \dfrac{1}{x^n}\right)$.

24. (i) If the complex numbers z_1, z_2, z_3 are connected by the relation

$$\dfrac{2}{z_1} = \dfrac{1}{z_2} + \dfrac{1}{z_3},$$

show that the points Z_1, Z_2, Z_3 representing them in an Argand diagram lie on a circle passing through the origin.

(ii) Express tan $(x+iy)$ in the form $u+iv$, where x, y, u, v are real.

If $\tan (x+iy) = \sin (p+iq)$, x, y, p, q real, prove that

$$\tan p \sinh 2y = \tanh q \sin 2x.$$

Section B. General

25. If $f(z) = u(x, y)+iv(x, y)$ is a differentiable function of $z = x+iy$ where x, y, u and v are real, establish the Cauchy–Riemann equations satisfied by u and v.

If, in some region, the differentiable function $f(z)$ is of the form

$$f(z) = u(x, y)+iv(x, y) = Xe^{iY},$$

where X and Y are real and Y is independent of x, prove that X is independent of y and determine all possible functions X, Y.

26. Explain what is meant by the statement that a function $f(z)$ of the complex variable z is differentiable at a point. If

$$f(z) = u(x, y)+iv(x, y)$$

where $z = x+iy$, and $u(x, y)$, $v(x, y)$, x, y are real, prove that

$$\frac{\partial u}{\partial x}+i\frac{\partial v}{\partial x} = f'(z) = \frac{\partial v}{\partial y}-i\frac{\partial u}{\partial y}$$

at any point at which $f(z)$ is differentiable.

Show that

(i) $$\frac{\partial^2 u}{\partial x^2}+\frac{\partial^2 u}{\partial y^2} = 0;$$

(ii) $$\left(\frac{\partial^2}{\partial x^2}+\frac{\partial^2}{\partial y^2}\right)|f(z)|^2 = 4|f'(z)|^2$$

in any region in which $f(z)$ is differentiable.

27. If $f(z)$ is regular when $0 < |z| \leqslant R$, and $zf(z)$ tends to a finite limit ζ as $z \to 0$, show that

$$\int_\gamma f(z)\, dz = 2\pi i\zeta,$$

where γ is the circle $|z| = R$ described in the positive (counter-clockwise) sense. Cauchy's theorem may be assumed.

If C is the circle $|z| = \varrho$ described positively, prove that if $\varrho > \frac{1}{2}$

$$\int_C \frac{e^{2\alpha z}}{4z^2+1}\, dz = \pi i \sin \alpha; \qquad \int_C \frac{2ze^{2\alpha z}}{4z^2+1}\, dz = \pi i \cos a.$$

What are the values of the integrals if $\varrho < \frac{1}{2}$?

28. (a) Express $1 + i \sqrt{3}$ in the form $r \exp (i\theta)$ and hence prove that $\left(\dfrac{1+i\sqrt{3}}{2}\right)^n$, where n is an integer, has only six distinct values.

Show that all but one of these values are roots of the equation

$$1 + z + z^2 + \ldots + z^5 = 0.$$

Hence find the roots of the equation

$$1 + w^2 + w^4 + \ldots + w^{10} = 0.$$

(b) Prove that, if $z = \exp (i\theta)$ and n is a positive integer, then

$$z^{2n} + z^{2n-2} + z^{2n-4} + \ldots + z^2 + 1 + z^{-2} + \ldots + z^{-2n} = \frac{\sin (2n+1)\theta}{\sin \theta}.$$

29. Test the following series for convergence:

(i) $$\sum_1^\infty z^n/n^2;$$

(ii) $$\sum_1^\infty e^{nz};$$

(iii) $$\sum_1^\infty n^2 i^n/(n^2 + 4).$$

30. If the sequence a_n of non-negative real numbers is decreasing, show that

$$a_{m+1} - a_{m-2} \leqslant a_{m+1} - a_{m+2} + \ldots + (-1)^{n-1} a_{m+n} \leqslant a_{m+1},$$

distinguishing between the cases of n odd and n even.

Deduce from the Cauchy principle of convergence that if $\lim a_n = 0$, then $\sum_1^\infty (-1)^n a_n$ converges. Writing r_m for $\sum_{m+1}^\infty (-1)^r a_r$, show that $|r_m| \leqslant a_{m+1}$.

By taking sufficiently many terms to ensure that the error in approximating to $\sum_1^\infty (-1)^r/r^4$ by a finite sum is less than 0.01, show that

$$-0.960 < \sum_1^\infty (-1)^r/r^4 < -0.939.$$

31. Prove that the series $\sum\limits_{n} n^{-r}$ is convergent if $p > 1$ and divergent if $p \leqslant 1$.

Determine the values of z (complex) and k (real) for which the following series converge *absolutely*:

(i) $\sum\limits_{n} n^k z^n$; (ii) $\sum\limits_{n} n^k e^{inz}$.

32. Explain what is meant by the statement that a function $f(z)$ of the complex variable z is differentiable at a point. If

$$f(z) = u(x, y) + iv(x, y),$$

where $z = x + iy$, and $u(x, y)$, $v(x, y)$, x, y are all real, prove that

$$\frac{\partial u}{\partial x} = \frac{\partial v}{\partial y}, \qquad \frac{\partial u}{\partial y} = -\frac{\partial v}{\partial x}$$

at any point at which $f(z)$ is differentiable.

Express $\sin z$ and the principal value of Log z ($z \neq 0$) in the form $u + iv$ and verify that the above equations hold in each case.

If

$$u = xy + \frac{x}{x^2 + y^2},$$

find the corresponding function v and express $u + iv$ explicitly in terms of z ($z \neq 0$).

33. Find the transformation of the form

$$w = \frac{az + b}{cz + d}$$

such that the points $z = i, 0, -i$ correspond to $w = 0, -i, 1 - i$ respectively. Show that the x-axis and the y-axis transform into a straight line and a circle respectively, and exhibit the region of the w-plane which corresponds to the first quadrant of the z-plane.

34. (i) Prove that if $f(z)$ is regular in and on a closed contour C, and ζ is a point within C, then

$$\int_C \frac{f(z)}{z - \zeta} \, dz = 2\pi i f(\zeta).$$

(ii) Prove that if $f(z)$ is regular and bounded in the whole of the z-plane, then f is a constant (Liouville's theorem). Deduce, or prove by any other method, that if

$$F(z) = z^m + a_1 z^{m-1} + \ldots + a^m,$$

where the a_r are complex numbers and m is a positive integer, then $F(z) = 0$ for at least one value of z.

35. (i) If $f(z) = U(x, y) + iV(x, y)$ is a differentiable function of z, prove that the curves $U =$ const. are orthogonal to the curves $V =$ const., and that the two families of curves

$$U^2 + V^2 = \text{const.} \quad \text{and} \quad U/V = \text{const.}$$

are likewise orthogonal. Discuss the case when $f(z) = 1/z$.

(ii) Show that, in the field of complex numbers, every algebraic equation

$$a_0 z^n + a_1 z^{n-1} + \ldots + a_n = 0 \quad (a_0 \neq 0,\ n \geqslant 1)$$

has a root.

36. State Cauchy's theorem concerning the integral of $f(z)$ round a simple closed contour C, and prove the theorem for the case when C is a triangle.

If Γ is the path from the point $z = 1$ to the point $z = -1$ along the upper half of the ellipse with equation $4x^2 + y^2 = 4$, prove that

$$\int_\Gamma \frac{dz}{(z+i)^2} = 1.$$

37. If $f(z)$ is a regular (analytic) function of $z\ (= x + iy)$ in a domain D, and if $f(z) = u(x, y) + iv(x, y)$, where u and v are real functions of x and y, deduce the Cauchy–Riemann equations

$$\frac{\partial u}{\partial x} = \frac{\partial v}{\partial y}, \quad \frac{\partial u}{\partial y} = -\frac{\partial v}{\partial x} \quad \text{in} \quad D.$$

State an additional condition that is required in order that the function $U(x, y) + iV(x, y)$ should be regular, given that U and V satisfy the Cauchy–Riemann equations in D.

Show that

(a) $x^2 + y^2 + 2ixy$ is not regular in any domain,

(b) $e^{3x} \cos 3y + ie^{3x} \sin 3y$ is regular everywhere,

(c) a line exists along which the expression in (a) satisfies the Cauchy–Riemann equations.

If $w = u + iv$ is regular, and $u = x^3 - 3xy^2 - 3x^2 + 3y^2$, determine v, and express w in terms of the single variable $z\ (= x + iy)$ given that $w = i$ when $z = 0$.

38. A bilinear transformation f is defined by $f(z) = \dfrac{2iz - 2}{2z - i}$.

(i) Determine the invariant points.

(ii) Find the point ζ for which the equation $f(z) = \zeta$ has no solution.

(iii) Show that the imaginary axis is mapped into itself.

(iv) State the condition satisfied by any circle that is transformed by f into a straight line.

(v) Determine the image of the disc $|z| < 1$, illustrating your answer by a sketch.

39. (a) State the Cauchy–Riemann equations in relation to the function

$$f(z) = u(x, y) + iv(x, y), \quad z = x + iy,$$

and explain their significance.

Show from first principles that $\dfrac{\partial}{\partial x} \sqrt{|xy|} = 0$ at the origin.

Verify that $f(z) = \sqrt{|xy|}$ satisfies the Cauchy–Riemann equations at the origin. Determine the limit of $[f(z) - f(0)]/z$ as z approaches 0 along the line $y = m^2 x$. Show that f is not regular at the origin.

(b) Show that the function

$$f(z) = e^x[(x \cos y - y \sin y) + i(y \cos y + x \sin y)], \quad (z = x + iy)$$

is regular over the whole complex plane.

40. State Cauchy's theorem.

Prove that if $f(z)$ is regular in and on a closed contour C and if a is a point within C, then

$$f(a) = \frac{1}{2\pi i} \int_C \frac{f(z)\,dz}{z - a}.$$

Indicate at what point in your proof you have needed $f(z)$ to be regular on the contour C.

Deduce that if $f(z)$ is regular in a domain S, its derivative at any interior point a is given by

$$f'(a) = \frac{1}{2\pi i} \int_C \frac{f(z)\,dz}{(z - a)^2},$$

where C is any simple closed contour lying within D and enclosing the point $z = a$.

41. Prove that under the transformation

$$w = \frac{az + b}{cz + d} \quad (ad \neq bc)$$

a circle γ in the z-plane is transformed into a circle or a straight line in the w-plane Characterize geometrically those circles γ whose transform is a straight line.

Given any point $z_0 \neq -d/c$ in the z-plane, prove that the circles through z_0 which are transformed into straight lines form a coaxal system.

42. (i) Defining $\sin z$, $\cos z$ by the equations

$$\sin z = \sum_0^\infty \frac{(-1)^n z^{2n+1}}{(2n+1)!}, \quad \cos z = \sum_0^\infty \frac{(-1)^n z^{2n}}{(2n)!},$$

prove that, for all z,

(a) $\dfrac{d}{dz}(\sin z) = \cos z,$

(b) $\sin 2z = 2 \sin z \cos z.$

(State carefully any theorem on power series which you use.)

(ii) Prove that the most general value of $\sin^{-1} 2$ is

$$(2k+\tfrac{1}{2})\pi \pm i \log(2+\sqrt{3}).$$

43. $u(x, y)$, $v(x, y)$ are functions of y, x whose partial derivatives exist in and on a triangle C. If

$$\left| \int_{C} (u+iv)\,(dx+idy) \right| = h,$$

establish the existence of a sequence of triangles $C, C_1, C_2, \ldots$, where the sides of C_r are half those of C_{r-1}, such that

$$\left| \int_{C_r} (u+iv)\,(dx+idy) \right| \geqslant h/4^r.$$

If $u+iv$ is an analytic function of $x+iy$, prove Cauchy's theorem, that $h = 0$.

44. Define exactly what is meant by the statement that $f(z)$ is an analytic function of z in a domain D.

If $u+iv$ is an analytic function of $x+iy$ in D, deduce the Cauchy–Riemann equations

$$\frac{\partial u}{\partial x} = \frac{\partial v}{\partial y}, \qquad \frac{\partial u}{\partial y} = -\frac{\partial v}{\partial x}$$

in D.

Show that functions u, v exist for which the Cauchy–Riemann equations are satisfied at a point, without $u+iv$ being analytic in any domain.

45. Prove that the transformation

$$z \to w = \frac{az+b}{cz+d}, \qquad (ad-bc \neq 0),$$

transforms the set of all lines and circles into itself.

46. (i) Explain what is meant by the statements that $f(z) = U(x, y)+iV(x, y)$ is differentiable (a) at a given point $z = \zeta$ and (b) in a given region R. Obtain the Cauchy–Riemann equations that must be satisfied by U, V if $f(z)$ is differentiable in R.

(ii) If Log w is the many-valued logarithmic function of w, prove that

$$\tan z = \frac{1}{2i} \, \mathrm{Log} \, \frac{1+iz}{1-iz}$$

and find explicitly all the values of $\tan^{-1}(i-2)$.

47. If a, b, c, d are any complex numbers such that $ad \neq bc$, prove that the transformation

$$w = (az+b)/(cz+d)$$

carries every circle in the z-plane into a circle or straight line in the w-plane. Show that the circles through the origin of the z-plane all transform to straight lines of the w-plane if $d = 0$.

If R is the upper half of the region given by $|z-\frac{1}{2}| \leqslant \frac{1}{2}$, find the region R' of the w-plane that corresponds to R under the transformation given by the equation

$$wz - z(1+i) + i = 0.$$

48. State and prove Taylor's theorem concerning the expansion in ascending powers of $z-a$ of a function $f(z)$ which is regular (single valued and differentiable) inside and on a circle C with centre a. [You may assume (i) the formula

$$\int_C \frac{f(z)}{z-\zeta} \, dz = 2\pi i f(\zeta)$$

for any point ζ inside C, and (ii) the term-by-term differentiability of a power series inside its circle of convergence.]

Find the expansion in ascending powers of $z-i$ of the function $f(z) = 1/(z^2-1)$, and state the radius of convergence of this series.

49. (i) If C is the triangular contour with vertices $1, 1+i, i$ taken in this order, prove that

$$\int_C \frac{dz}{z^3-i} = \frac{\pi}{3}(\sqrt{3}+i).$$

(ii) Prove that, if $a > 0$,

$$\int_0^\infty \frac{\cos x \, dx}{x^4+a^4} = \frac{\pi}{2a^3} e^{-a/\sqrt{2}} \sin(a/\sqrt{2}+\pi/4).$$

50. State Cauchy's integral theorem, and deduce the integral formula

$$f(\zeta) = \frac{1}{2\pi i} \int_C \frac{f(z)\, dz}{z-\zeta},$$

where ζ is a point inside the closed contour C, the conditions for Cauchy's theorem being satisfied.

Give an outline of the derivation, from this formula, of the Taylor expansion of $f(z)$ in a neighbourhood of the point ζ.

Find the coefficient of $(z+1)^n$ in the expansion of $(2-3z+z^2)^{-1}$ in a series of positive powers of $z+1$. For what region of the z-plane is this expansion valid?

51. (i) Evaluate the contour integral

$$\int_C \frac{dz}{z^4+1},$$

where C is the rectangle with vertices $\pm i, 1\pm i$.

(ii) Use contour integration to prove that, if $a > 0$,

$$\int_0^\infty \frac{dx}{(x^2+a^2)^3} = \frac{3\pi}{16a^5}.$$

52. If a, b, c and d are complex numbers such that $ad \neq bc$ prove that the transformation $w = \dfrac{az+b}{cz+d}$ carries every circle in the z-plane into a circle or straight line in the w-plane. Show that if $d = 0$, the circles through the origin of the z-plane transform into straight lines of the w-plane.

If D is the upper half of the region $|z-\frac{1}{2}| \geqslant \frac{1}{2}$ find the region D' of the w-plane which corresponds to D under the transformation given by

$$wz-z(1+i)+i = 0.$$

53. (a) Under the transformation $w = \dfrac{z-i}{z+i}$, determine the images in the w-plane of:

(i) the interior of the unit circle $|z| = 1$;

(ii) the half-plane $\mathcal{R}(z) > 0$;

(iii) the line segment joining the points $1, -i$.

(b) Under the transformation $w = e^z$ determine the images in the w-plane of

(i) the real axis;

(ii) the area bounded by the real axis and the line $y = \pi$. Illustrate by diagrams.

54. Using the methods of contour integration, evaluate

(i) $\displaystyle\int_0^{2\pi} \frac{d\theta}{1+a\cos\theta}$, where $-1 < a < 1$, stating the reason for the restriction on a;

(ii) $\displaystyle\int_0^\infty \frac{\cos x\,dx}{(1+x^2)^2}.$

55. (a) Express $\dfrac{1}{(z^2-4)(z^2+1)}$ as the sum of a Taylor or Laurent series (whichever is appropriate) for each of the regions:

(i) $|z| < 1$;   (ii) $1 < |z| < 2$;   (iii) $2 < |z|$.

(b) Determine the principal part, and the first term of the regular part, of the Laurent expansion of $\dfrac{1}{z(e^z-1)}$ in powers of z.

56. If $f(z)$ is a regular function in and on a closed contour C, Cauchy's theorem states that

$$\int_C f(z)\, dz = 0.$$

Assuming that the theorem is true whenever f has the form $f(z) = az+b$, where a and b are constants, prove it *either* when C is a triangle *or* when C is a square and f is not linear.

Evaluate

$$\int_L (z^2-z)\, dz,$$

where L is the path consisting of the two straight segments OA, AP, where O is the origin, A the point $2+i$ and P the point 3.

57. Using the methods of contour integration, evaluate

(i) $\displaystyle\int_0^{2\pi} \dfrac{\cos 2\theta}{5+4\cos\theta}\, d\theta$;

(ii) $\displaystyle\int_0^{\infty} \dfrac{\cos x\, dx}{x^2+9}$.

58. (i) Show that if z, a and ζ are distinct complex numbers, then

$$\frac{1}{z-\zeta} - \frac{1}{z-a} + \frac{\zeta-a}{(z-a)^2} + \cdots + \frac{(\zeta-a)^{n-1}}{(z-a)^n} + \frac{(\zeta-a)^n}{(z-a)^n}\frac{1}{(z-\zeta)}.$$

Deduce that if $F(z)$ is regular in $|z-a| \leqslant \varrho$, and $|\zeta-a| < \varrho' < \varrho$, then

$$F(\zeta) = F(a) + \sum_{r=1}^{n-1} a_r(\zeta-a)^r + R_n,$$

where $R_n \to 0$ as $n \to \infty$, and a_r is defined by the formula

$$a_r = \frac{1}{2\pi i} \int_{C'} \frac{F(z)}{(z-a)^{r+1}}\, dz,$$

C' being the circle $|z-a| = \varrho'$ described positively.

(ii) Express

$$\frac{5z}{(2z-1)(z+2)}$$

as the sum of a power series in z and state for what values of z the expansion is valid.

59. By means of the substitution $z = e^{i\theta}$ and integration round the unit circle prove that

$$\int_0^\pi \frac{d\theta}{a+b\cos\theta} = \frac{\pi}{\sqrt{(a^2-b^2)}} \quad (a > b > 0).$$

Hence or otherwise deduce that

$$\int_0^\pi \frac{d\theta}{(a+b\cos\theta)^2} = \frac{\pi a}{(a^2-b^2)^{3/2}} \cdot$$

60. Show, by contour integration, that

$$\int_0^\infty \frac{\cos mx}{(a^2+x^2)^2}\,dx = \frac{\pi}{4a^3}(1+am)e^{-am},$$

where $a > 0$ and $m > 0$.

REFERENCES

1. ALLEN, R. G. D., *Basic Mathematics*, Macmillan, 1962.
2. ESTERMANN, T., *Complex Numbers and Functions*, Athlone, 1962.
3. HARDY, G. H., *Pure Mathematics*, 9th edn., Cambridge, 1944.
4. HARDY, G. H., *Divergent Series*, Oxford, 1949.
5. PENNISI, L. L., *Elements of Complex Variables*, Holt, Rinehart & Winston, 1967.
6. PHILLIPS, E. G., *Functions of a Complex Variable*, 6th edn., Oliver & Boyd, 1949.
7. TITCHMARSH, E. C., *The Theory of Functions*, 2nd edn., Oxford, 1939.

ANSWERS TO SELECTED EXERCISES AND MISCELLANEOUS EXERCISES

Answers to Selected Exercises

Chapter 1

1.3.1. $\tan \theta = -1, 4/3, 0, \infty, -4/3, -1, 0, 1, -1, 4/3, -\infty$.
modulus $= 3\sqrt{2}, 5, 3, 3, 5, 3\sqrt{2}, 3, 3\sqrt{2}, 3\sqrt{2}, 5, 3$.

1.3.2. (a) On same ray from origin; (b) each on ray containing conjugate of other; (c) (i) on same line through origin, (ii) on same ray through origin, (iii) on same line through origin. [In (c) (i) and (iii) the points are separated by the origin.]π, 0.

1.3.3. Modulus r; amplitudes $\theta + n\pi/3$ ($n = 1, 2, 3, 4, 5$).

1.4.1. (a) $\sqrt{2}, \pi/4$; $5\sqrt{2}, -\pi/4$; $13, \pi - \tan^{-1} 12/5$; $4, \pi$; $2, 2\pi/3$; $4, -\pi/2$.
   (b) (i) $(0, 5)$; (ii) $(-1, 1)$; (iii) $(-6, 0)$; (iv) $(2\pi, 0)$; (v) $(\sqrt{3}, 3)$.

1.5.2. $(-1, 7)$. (a) $(-6, 14)$; (b) $(-3, 7)$; (c) $(3, -7)$; (d) $(0, 0)$.

1.5.3. $(1, 1)$.

1.5.5. $(\frac{1}{2}, \frac{1}{4})$.

1.6.2. Either $k = 0$ or $z = 0$.

1.7.1. $(\theta \pm \pi)$.

1.7.2. (a) $(4, 4)$; (b) $z = (5, 2)$, $w = (3, 4)$; (c) $z = (2, 1)$, $w = (0, 3)$.

1.8.1. (a) $(0, 25)$; (b) $(-2, 26)$; (c) $(0, 2)$; (d) $(-85, 0)$; (e) $(-64, 0)$; (f) $(-22, 48)$; (g) $(-55, 40)$.

1.8.2. $(-2, -1)$.

1.8.3. $x = a/(a^2+b^2)$, $y = -b/(a^2+b^2)$.

1.8.5. (a) $z = \pm(1/\sqrt{2}, 1/\sqrt{2})$; (b) $(0, 1)$.

1.8.6. $(3, 2)$ and $(-3, -2)$.

1.8.7. $(2, 0)$; $(1, 0)$. $(-1, 1)$, $(-1, -1)$. $b \pm \sqrt{(-c)}$.

1.9.2. (a) $(3, 1)$; (b) $(4/5, -3/5)$; (c) $(-5, 0)$; (d) $(x, -y)$; (e) $[x/(x^2+y^2), -y/(x^2+y^2)]$.

1.10.1. (a) 25; (b) $1+76i$; (c) $3\frac{1}{2}+1\frac{1}{2}i$; (d) 1; (e) 1; (f) i.

1.10.2. (a) $3i, 2i$; (b) $1, i, -1, -i$; (c) $1, (-1\pm\sqrt{3})/2$.

1.11.1. (a) $\frac{1}{2}(1-i)$; (b) $(4+3i)/25$; (c) $(2+i)/10$; (d) $(2-i)/10$; (e) -1; (f) $-i$; (g) $(7-5i)/74$.

1.11.2. (a) $i/2$; (b) $4/5$.

1.11.3. (a) $3, \pi$; (b) $1, -\pi/2$; (c) $\sqrt{2}, 3\pi/4$; (d) $2, -\pi/4$; (e) $2, -2\pi/3$.

1.12.1. 1.

1.12.2. $1+i, 3+i$.

1.12.3. $0-i1$.

1.13.4. (a) The real axis; (b) the imaginary axis; (c) circle, centre 0, radius r; (d) circle, centre a, radius b; (e) perpendicular bisector of join of a, b; (f) straight line.

1.13.5. $(z^2+\bar{z}^2)(1/a^2-1/b^2)+2z\bar{z}(1/a^2+1/b^2) = 4$.

1.14.4. E.g. $(z_1-z_2)/(z_3-z_2)$ is real. [Consider amp (z_1-z_2), amp (z_3-z_2).]

1.15.1. $z = -1, \frac{1}{2}\pm i \sqrt{(3)/2}$.

1.15.2. $\pm(1+i)/\sqrt{2}, \pm(1-i)/\sqrt{2}, \pm 2i, \pm 2^{1/4} (\cos \pi/8 + i \sin \pi/8)$.

1.15.3. (a) $z^2-(1+i)z+i = 0$; (b) $z^2-3iz-2 = 0$; (c) $z^2-2z+5 = 0$; (d) $z^2-2\alpha z+(\alpha^2+\beta^2) = 0$. [(c) and (d) have real coefficients.]

1.15.4. $z = \cos (2n+1)\pi/5 + i \sin (2n+1)\pi/5$ $(n = 0, 1, 2, 3, 4)$.

1.15.5. $\prod_{r=1}^{n-1} [x^2 - 2 \cos (2r+1)\pi/2n+1]$.

1.15.6. $2^{3/4} (\cos \theta + i \sin \theta)$, where $\theta = (3+6n)\pi/8$ $(n = 0, 1, 2, 3)$.

1.15.8. $\cos 5\theta = 16 \cos^5 \theta - 20 \cos^3 \theta + 5 \cos \theta$.
$\sin 5\theta = 16 \sin^5 \theta - 20 \sin^3 \theta + 5 \sin \theta$.

1.15.9. $(\cos 6\theta + 6 \cos 4\theta + 15 \cos 2\theta + 10)/32$.

1.15.10. $z = \cos r\pi/3 + i \sin r\pi/3$. $(r = 1, 2, 3, 4, 5)$.

Chapter 2

2.1.3. (i) Open; (ii) closed; (iii) open; (iv) half-open.

2.2.1. (a) l.u.b. 2, g.l.b. 0; (b) l.u.b. 100, g.l.b. -100; (c) l.u.b. 100; (d) l.u.b. 10, g.l.b. 1; (e) l.u.b. 1, g.l.b. 0; (f) g.l.b. 2; (g) (i) l.u.b. 1, g.l.b. 0; (ii) l.u.b. 1, g.l.b. 0; (iii) l.u.b. 1, g.l.b. -1; (iv) g.l.b. -2.

2.2.2. (a) 1, not a member; (b) all points in set; members; (c) none; (d) none; (e) 0, 1, not members; all members of set; (f) all members of set.

2.4.1. (a) Yes; (b) yes; (c) no; (d) yes; (e) yes; (f) no; (g) no; (h) yes; (i) no.

2.5.1. (a) No; (b) no; (c) yes; (d) no; (e) no; (f) yes; (g) no.

2.6.1. (a) $1+i$; (b) all; (d) 0; (d) (1, 0).

2.7.1. 2.4.1: (c), (d); 2.5.1: (a), (b), (f), (g).

2.10.1. (a) 1000; (b) 1000; (c) 1000; (d) 1999.

2.11.4. (a) 1; (b) dgt; (c) 0; (d) 2; (e) $i\pi$; (f) dgt.

2.13.1. (a) $4\frac{2}{3}$; (b) $4\frac{1}{2}$; (c) $\sqrt{5}$; (d) $\sqrt{2}$; (e) 5.

2.14.1. (a) $1+i$; (b) dgt; (d) dgt; (e) 0.

2.16.2. 1.

2.17.2. (a) $1\frac{1}{9}$; (b) 3; (c) 3; (d) 0.

Chapter 3

3.3.2. $\frac{1}{2}[1-1/(2n+1)]$; 1.

3.4.4. (a) dgt; (b) cgt; (c) cgt; (d) cgt; (e) cgt.

3.5.2. (a) cgt; (b) cgt; (c) cgt; (d) cgt; (e) dgt.

3.5.3. $-1 < x < 1$.

3.6.3. Upper limit $7+\frac{1}{2} \sqrt{3}$; lower limit 4.

3.7.2. $|x| < 1$.
3.7.3. $|x| < 4$.
3.8.2. cgt, $p > 1$.
3.9.2. (a) cgt; (b) dgt; (c) cgt; (d) cgt; (e) dgt; (f) dgt; (g) cgt.
8.10.1. (a) cgt; (b) cgt; (c) cgt.
3.10.2. (a) $-1 \leqslant x < 1$; (b) all; (c) $-1 \leqslant x \leqslant 1$; (d) $-1 \leqslant x \leqslant 1$.
3.10.3. (a) $|z| < 1$; (b) $|z| \leqslant 1$. If $A = \pi$, cgt for all z.
3.10.4. When $|z| = 1$, (a) is dgt, (b) is dgt.
3.11.3. (a) 2; (b) 1; (c) 2; (d) e.
3.11.4. (a) dgt, cgt, cgt, cgt.
     (b) cgt, cgt, cgt, cgt.
     (c) dgt, dgt, dgt, dgt.
3.11.5. (a) dgt; (b) cgt; (c) cgt; (d) dgt.
3.11.6. The half-plane $x < 0$.

Chapter 4

4.1.1. (i) 0, 7, 8; (ii) $3+4i, 2-2i, 1-4i$; (iii) $5-12i, 3+4i, -3+4i$; (iv) $(5-13i)/12$, $(18+26i)/25, (-18+26i)/40$.
4.1.2. (a) $u = x^2-3xy^2$, $v = 3x^2y-y^3$.
     (b) $u = (x^2+y^2-1)/(x^2+y^2+2x+1)$, $v = 2y/(x^2+y^2+2x+1)$.
     (c) $u = x(x^2+y^2)$, $v = y/(x^2+y^2)$.
     (d) $u = x^2-y^2+x+1$, $v = 2xy+y$.
4.2.1. $z = \pm 2i$.
4.2.3. (a) 0.02; (b) 0.4 approx.; (c) 141.4 approx.
4.2.4. (a) 1; (b) e.g. $|\delta z| < 0.2$; (c) $|\delta z| < 1/42$, say.
4.5.2. $\varepsilon = 0.2, \eta = -0.1, \varepsilon' = 0.1, \eta' = 0.2$.
4.6.1. (b) $z = \pm 2i$; (c) $z = 5$.
4.6.3. e^{mx} (cos $my + i$ sin my), $(x^2+y^2)/(3x-3iy)$, $(1-z^4)/(1+z^4)$.
4.8.1. (a) $x^2-y^2 = c$, $xy = k$.
     (b) $x^2+y^2 = cx$, $x^2+y^2+ky = 0$.
     (c) $x^2+y^2+2x+1 = c(x+1)$, $x^2+y^2+2x+1 = ky$.
4.8.2. (a) $u = \sin x \cosh y$; (b) $v = e^{x^2-3xy^2} \sin (3x^2y-y^3)$; (c) $u = (x+1)/(x^2+y^2+2x+1)$; (d) $v = -x$; (e) $v = \tan^{-1}(y/x)$; concentric circles and rays through common centre; (f) $v = y^2-x^2$; (g) $u = x^3-3x^2y$.

Chapter 5

5.1.1. (a) $z = 0, \pm 2i$; (c) $3z^2+4$.
5.2.1. $z = \pm 2i$. $dw/dz = 16/25$; $\frac{1}{2}$; $16i/9$; $16 (9-2i)/17$.
5.3.1. 4.11.
5.4.3. $z+z^3/3+2z^5/15$.
5.5.2. $\pm i \log (2+\sqrt{3})$; $-i \log (1+\sqrt{2})$.
5.5.4. $32^{1/4} \exp (-i\pi)/8$; $32^{1/4} \exp (7\pi i/8)$.

5.5.5. (a) 6; (b) -1; (c) -1; (d) i; (e) $-i$; (f) $1+i$.

5.6.1. $\exp(2n\pi i)$; $\exp(2n+1)\pi i$; $\exp(2n+\tfrac{1}{2})\pi i$; $\exp(2n-\tfrac{1}{2})\pi i$.

$\exp 0$; $\exp \pi i/2$; $\exp \pi i$; $\exp 3\pi i/2 \quad (=1, i, -1, -i)$.

$\exp \pi i/4$; $\exp 3\pi i/4$; $\exp(-\pi i/4)$; $\exp(-3\pi i/4) \quad (=\pm 1/\sqrt{2} \pm i/\sqrt{2})$.

$\exp \pi i/8$; $\exp 5\pi i/8$; $\exp(-3\pi i/8)$; $\exp(-7\pi i/8)$.

$\exp(-\pi i/8)$; $\exp 3\pi i/8$; $\exp(-5\pi i/8)$; $\exp 7\pi i/8$.

5.7.1. $\cosh z = 1 + z^2/2! + z^4/4! + \ldots + z^{2n}/(2n)! + \ldots$

$\sinh z = z + z^3/3! + z^5/5! + \ldots + z^{2n+1}/(2n+1)! + \ldots$

5.7.5. $z = 2$.

5.7.6. $z = \log(3 + \sqrt{10})$.

5.7.8. $u = \operatorname{sh} x \operatorname{ch} x/(\cos^2 y \operatorname{ch}^2 x + \sin^2 y \operatorname{sh}^2 x)$.

$v = \sin x \cos y/(\cos^2 y \operatorname{ch}^2 x + \sin^2 y \operatorname{sh}^2 x)$.

5.8.1. (a) $(2n+\tfrac{1}{2})\pi i$; (b) $(2n+1)\pi i$; (c) $\tfrac{1}{2}\log 2 + 2n\pi i + i\pi/4$; (d) $\log 2 + \pi i(2n-\tfrac{1}{6})$.

5.8.2. $\log(1+z) = z - z^2/2 + z^3/3 - \ldots \operatorname{cgt} |z| < 1$.

$\log(2+z) = \log 2 + z/2 - \tfrac{1}{2}(z/2)^2 + \tfrac{1}{3}(z/2)^3 - \ldots \operatorname{cgt} |z| < 2$.

$\log(i+2z) = i\pi/2 - 2iz + 2^2 z^2/2 + i2^3 z^3/3 - 2^4 z^4/4 + \ldots \operatorname{cgt} |z| < \tfrac{1}{2}$.

5.8.4. (a) $z = 2$; $-1 \pm \sqrt{3}$; (b) $z = \pm i$; (c) $z = 0$; (d) $z = 2n\pi$ $(n = 0, \pm 1, \pm 2, \pm 3, \pm 4 \ldots)$;

5.9.3. (a) $\exp(2n-\tfrac{1}{2})\pi$ $(n = 0, \pm 1, \pm 2, \ldots)$; (b) $\exp(2n-1)\pi$ $(n = 0, \pm 1, \pm 2, \ldots)$; (c) $\exp(2n+1)\pi$ $(n = 0, \pm 1, \ldots)$; (d) $\exp 3 (\cos 4 + i \sin 4)$; (e) $\exp(\tfrac{1}{2}i \log 2 + 2n\pi - \pi/4)$; (f) $\exp(4n-2)\pi/3(\cos \log 2 + i \sin \log 2)$.

5.10.1. 0, r even; $(-1)^{(1/2)\,(r-1)}$, r odd.

5.12.1. $z + z^3/3! + z^5/5! + \ldots$.

5.12.2. (a) $z + z^3/3 + z^5/5 + \ldots$; (b) $-z - z^2/2 - z^3/3 - z^4/4 - \ldots$; (c) $1 + z \log a + (z \log a)^2/2! + (z \log a)^3/3! + \ldots$.

Chapter 6

6.1.1. (a) $Z(m+i) + \bar{z}(m-i) + 2c = 0$; (b) $z(b-ai) + z(b+ai) = 2ab$; (c) $z[y_2-y_1 + i(x_2-x_1)] = 2(x_1 y_2 - y_1 x_2)$; (d) $z(\bar{z}_2 - \bar{z}_1) - \bar{z}(z_2 - z_1) + (\bar{z}_1 z_2 - z_1 \bar{z}_2) = 0$; (e) $z(i+m) - \bar{z}(i-m) + 2(y_1 - mx_1) = 0$; (f) $z = \bar{z}$; (g) $z + \bar{z} = 0$; (h) $z - \bar{z} = ik$ (k real), $z \mid \bar{z} = h$ (h real); (i) $z(m+i) + \bar{z}(m-i) = 0$; (j) $z \exp(-i\theta) + \bar{z} \exp(i\theta) = 2p$.

6.1.2. (a) $b\bar{a} - a\bar{b} = 0$, (b) $a\bar{b} + b\bar{a} = 0$.

6.1.3. (a) $az + \bar{a}\bar{z} = a(p+iq) + \bar{a}(p-iq)$; (b) $az - \bar{a}\bar{z} = a(p+iq) - \bar{a}(p-iq)$.

6.1.4. $z(a-c-ib+id) + \bar{z}(a-c+ib-id) = a^2 + b^2 - c^2 - d^2$.

6.1.5. (a) The point $z = -3i$; (b) the points $z = \pm ia$; (c) the point $z = -b/a$.

6.2.1. $\operatorname{amp}(z-1-i) = \pi/4, 5\pi/4. 5\pi/4$. (The line may be written $\operatorname{amp} z = \pi/4, 5\pi/4$.)

6.2.2. (a) $\operatorname{amp}(z-ci) = \tan^{-1} m$, $\pi + \tan^{-1} m$; (b) $\operatorname{amp}(z-a) = -\tan^{-1} b/a$, $\pi - \tan^{-1} b/a$; (c) $\operatorname{amp}(z-x_1-iy_1) = \tan^{-1}(y_2-y_1)/(x_2-x_1)$, $\pi + \tan^{-1}(y_2-y_1)/(x_2-x_1)$; (d) $\operatorname{amp}(z-z_1) = \operatorname{amp}(z_2-z_1)$, $\pi + \operatorname{amp}(z_2-z_1)$; (e) $\operatorname{amp}(z-x_1 - iy_1) = \tan^{-1} m$, $\pi + \tan^{-1} m$; (f) $\operatorname{amp} z = 0$, π; (g) $\operatorname{amp} z = \pi/2$, $-\pi/2$; (h) $\operatorname{amp}(z-ib) = 0$, π; $\operatorname{amp}(z-a) = \pm\pi/2(a, b \text{ real})$; (i) $\operatorname{amp} z = \tan^{-1} m$, $\pi + \tan^{-1} m$.

6.2.3. (f) $|z-a| = |z-\bar{a}|$, a not real (e.g. $|z-i| = |z+i|$);
(g) $|z-a| = |z+\bar{a}|$, e.g. $|z-1| = |z+1|$;
(h) e.g. $|z-ia| = |z-ib|$, a, b, real; $|z-a| = |z-b|$, a, b real;
(j) e.g. $|z| = |z-2p\exp(i\alpha)|$.

6.2.4. $Z = [-\bar{a}(p-iq)-c]/a = (-\bar{a}\bar{w}-c)/a$.

6.3.3. amp $[(z-2i)/z] = \pi/2$; amp $[(2i-z)/z] = \pi/2$.

6.3.4. amp $(z+i)/(z-i) = \pi/2$.

6.3.5. (d) e.g. $(z_1-z_2)/(z_3-z_2)$ is real for straight line;
(e) $|(z-z_1)(z-z_2)| = |(z_3-z_1)(z_3-z_2)|$.

6.3.7. e.g. $[(z_1-z_2)(z_3-z_4)/(z_1-z_4)(z_3-z_2)]$ real.

Chapter 7

7.1.2. $z' = z+\bar{a}$. Centre of circle translated to 0.

7.2.1. (a) All points on line; (b) given line and all lines perpendicular to it; (c) $-(p-iq+c)$; (d) $-\bar{a}(p-iq)/a$; (e) $-[\bar{a}(p-iq)+c]/a$.

7.5.1. $w = i\bar{z}+k(1+i)$.

7.5.2. $w = -(\bar{a}\bar{z}+c)/a+i\bar{a}d/|a|$.

7.7.4. $y = 4ax(x^2+y^2)$.

7.8.2. Infinity mapped on to (a) 0; (b) ∞; (c) a/c; (d) 0. Points mapped on to infinity are (a) all points $a+ib$, where $a \to -\infty$; (b) ∞; (c) $-d/c$; (d) $z = \pm i$.

Chapter 8

8.2.2. $z = [a-d\pm\sqrt{(a^2+d^2-2ad-4bc)}]/2c$.

8.3.2. $|(z-\tfrac{1}{2})/(z-2)| = \tfrac{1}{2}$ transforms to $|(w+3/5+4i/5)/(w-3/5+4i/5)| = 1$ (imaginary axis).

8.3.4. (a) $|z| = 1$; (b) circle (line) passes through $z = -i$; (c) the point $w = i$.

8.4.1. (a) (i) Upper half-plane; (ii) lower half-plane; (iii) interior of circle $4w\bar{w}-5(w+\bar{w})+4 = 0$; (iv) upper half-plane, $|c| > 1$; lower half-plane, $|c| < 1$. Point 1, $c = 1$. Point -1, $c = -1$.
(b) (i) Exterior of unit circle; (ii) exterior of circle centre $\tfrac{1}{2}$, radius $\tfrac{1}{2}$; (iii) lower half-plane; (iv) circle centre $\tfrac{1}{2}(c+1/c)$, radius $\tfrac{1}{2}(c-1/c)$ ($c \neq 0$, 1, -1).
(c) (i) Half-plane $\mathcal{R}(w) < 0$; (ii) interior of circle centre $-\tfrac{1}{3}$, radius $\tfrac{2}{3}$; (iii) interior of circle centre $\tfrac{2}{3}$, radius $\tfrac{2}{3}$; (iv) ($c \neq 1$) interior of unit circle, $c > 1$; exterior, $c < 1$.
(d) (i) Circle centre $-i$ radius $\sqrt{(2)}$; (ii) circle centre $\tfrac{1}{2}(1-i)$, radius $1/\sqrt{2}$; (iii) circle centre $(5+3i)/4$, radius $\tfrac{1}{2}\sqrt{30}$; (iv) circle centre $(1+c^2)/2c+i(1-c^2)/2c$, radius $(1-c^2)\sqrt{2}/2c$.
(e) (i) Circle centre 1, radius 2; (ii) imaginary axis; (iii) circle amp $[(w-5/4+3i/4)/(w-\tfrac{1}{2})] = \pi/4$, $-3\pi/4$; (iv) circle amp $\{[w-(1+2c)/(2+c)]/(w-1/c)\} = \pm\pi/2-$amp $(c+2)/c$.

8.4.3. Intersection of interior of $u^2+v^2-2\tfrac{1}{2}u+v+1\tfrac{1}{2} = 0$ with exterior of $u^2+v^2+v-1 = 0$.

8.5.1. $w = \exp(i\theta)(z-a)/(\bar{z}-\bar{a})$; $\mathcal{I}(a) > 0$, θ real.

8.5.2. $w = (az+b)/(cz+d)$; a, b, c, d real; $ad-bc > 0$.
8.5.3. $w = R \exp(i\theta)(z+a)/(z+\bar{a})$; $\mathfrak{J}(a) > 0$.
8.6.1. $w = R \exp(i\theta)(z-ar)/(\bar{a}z-1)$; $\mathfrak{J}(a) < 0$.
8.6.2. $w = i(a+1)(z-1)/(az-1)$; a real, $-1 < a < 1$.
8.8.1. (a) the upper half of the interior of $r = 2(1+\cos\theta)$; (b) the interior of the rectangle $v = 2a$, $v = 2b$, $u = c$, $u = d$; (c) the real axis from -1 to $+1$ and two parabolic arcs; (d) the part of the complex plane exterior to the circle $|z| = 4$.
8.8.2. Strips bounded by confocal parabolas with the real axis as axis.
8.8.3. $z = 1, (1 \pm i\sqrt{7})/2$.
8.9.1. Interior of lower semicircle of unit circle.
8.9.3. Part of annulus, radii 1, e. Complete annulus.
8.12.4. Circles amp $\{(z-1)/(z+1)\} = \pm\pi/4, \pm3\pi/4$.

Chapter 9

9.1.1. (a) $az^3/3+bz^2/2+cz+d$; (b) $\frac{1}{2}\exp(z^2)+c$; (c) $\frac{1}{2}z\exp(2z)-\exp(2z)/4+c$;
(d) $\log z+c$ ($z \neq 0$); (e) $3z+5\log(z-1)+c$ ($z \neq 1$); (f) $\{\sin(5z)\}/5$.
9.2.1. (i) (a) 0; (b) 0; (c) i; (ii) (a) 0; (b) 0; (c) $(2i-2)/3$; (iii) (a) 0; (b) $2+2i$; (c) $(14-5i)/15$; (iv) (a) $\pi(i-1)$; (b) $i-1$; (c) $(5+7i)/6$; (v) (a) $-i\pi$; (b) $-i$; (c) $-\frac{1}{2}+7i/3$; (vi) (a) 0, n even, $2\pi in(n-2)\ldots3, 1/(n+1)(n-1)\ldots2$, n odd; (b) 0; (c) $1/(n+1)+2i/(n+2)$; (vii) (a) 0; (b) 0; (c) $k(1+i)$.
9.4.1. (a) e^b-e^a; (b) $[\cos(pb+q)-\cos(pa+q)]/p$; (c) $3(b-a)+5\log[(b-1)/(a-1)]+2n\pi i$, where n depends on contour (see 9.4.2); (d) $(1/a-1/b)$. (Contour not through 0.)
9.4.2. $2\pi i$.
9.5.1. (These answers are not unique.) (a) 20; (b) $2\pi R/(R^2-1)$; (c) $4e$, $2\pi e$; (d) $2\pi R^2/(|a|R^2-|b|R-|c|)$; (i) 1; (ii) $\exp R$.

Chapter 10

10.2.2. (a) $z-z_0$; (b) $z^2+zz_0-2z_0^2$; (c) $(z-z_0)/zz_0^2$.
10.4.1. (i) $2\pi i$; (ii) $2\pi i$, 0; (iii) $2\pi i$; (a) 0; (b) $2\pi i$.
10.5.2. $-\pi(e-e^{-1})$.
10.6.2. $4\pi i$.
10.7.1. (i) 0; (ii) 0; (iii) 0.
10.8.2. (a) not defined at $z = -a^2$; (b) at $z = \pm ib$; (c) at $z = 0$; (d) at $z = (2n+1)\pi i/2$.

Chapter 11

11.1.1. $1/(1-z)^2-z^n(1+n-nz)/(1-z)^2$.

$$1/(z-Z)^2 = \sum_1^n r(Z-a)^r/(Z-a)^{r+2} + \frac{(Z-a)^n(z-a+nz-nZ)}{(z-a)^{n+1}(z-Z)^2}.$$

11.1.2. $1/(z-Z) = -\sum\limits_{1}^{n} (z-a)^{r-1}/(Z-a)^r + (z-a)^n/(Z-a)^n (Z-z).$

11.2.2. (i) $\sum\limits_{0}^{\infty} (-1)^r (z-a)^r/a^{r+1}$ $(a \neq 0, R = |a|).$

    (ii) $\sum\limits_{0}^{\infty} (-1)^r (z-a)^r/(a-b)^{r+1}$ $(a \neq b, R = |b-a|).$

    (iii) $e^a \sum\limits_{0}^{\infty} (z-a)^r/r!$ $(R = \infty).$

    (iv) $\sum\limits_{1}^{\infty} (-1)^{r+1} (z-a)^r/ra$ $(R = |a|).$

    (v) $\sin ka \sum\limits_{0}^{\infty} (-1)^r k^{2r} (z-a)^{2r}/(2r)!$

        $+\cos ka \sum\limits_{1}^{\infty} (-1)^{r+1} k^{2r-1}(z-a)^{2r-1}/(2r-1)!$ $(R = \infty).$

    (vi) $\sum\limits_{0}^{\infty} a^{n-r} \binom{n}{r} (z-a)^r$ $(R = a).$

11.2.3. (i) $\pi/4 + \frac{1}{2}(z-1) - (z-1)^2/4;$
    (ii) $\frac{1}{2}\log 13 + i(\pi - \tan^{-1}\frac{2}{3}) - (3+2i)t/13 - (3+2i)^2 t^2/338;$
    (iii) $1 + z + z^2/2.$

11.3.1. 5.

11.3.2. (i) (a) 1; (b) $\surd(26)$; (c) 1; (ii) 1; (iii) 1.

11.4.1. (i) (a) $\sum\limits_{0}^{\infty} z^{2r+1}/(2r+1)!$ $(R = \infty);$ (b) $\sum\limits_{0}^{\infty} -(z-\pi i)^{2r+1}/(2r+1)!$

    (ii) (a) $\sum\limits_{0}^{\infty} (-\frac{1}{5} \times 3^{r+1} + (-1)^{r+1}/5 \times 2^{r+1})t^r$ $[t = (z+1)];$

     (b) $\sum\limits_{0}^{\infty} (-\frac{1}{5} \times 4^{r+1} + (-1)^{r+1}/5)t^r$ $[t = (z+2)];$

     (c) $\sum\limits_{0}^{\infty} (-1)^r (1/3^{r+1} - 1/8^{r+1})t^r/5$ $[t = (z-5)].$

11.4.2. $1 + x^2/2 + 5x^4/24.$

11.5.1. (a) $\sum\limits_{0}^{\infty} (-1)^r z^{4r+2}/(2r+1)!$

    (b) $\sum\limits_{0}^{\infty} (-1)^r 2^{2r+1} z^{2r+1}/(2r+1)!$

    (c) $\sum\limits_{1}^{\infty} (-1)^{r+1} i^r z^r/r.$

    (d) $\sum\limits_{0}^{\infty} c^{2r+1} z^{2r+1}/(2r+1)!$

    (e) $\sum\limits_{1}^{\infty} (-1)^r z^{2r}.$

(f) $\sum_1^\infty (-1)^{r+1} r z^{r+1}$.

(g) $\sum_0^\infty (-1)^r a^{2r+1} z^{2r+1}/(2r+1)$.

(h) $\sum_1^\infty (-1)^{r+1} z^{2r}/r$,

11.6.1. $|J_n| \leqslant 2/|z|^n (|z|-1)$.

11.6.2. (a) $\sum_1^\infty [(-1)^r 2^r - 1/2^r] z^r/5 \quad (R = \tfrac12)$.

(b) $\sum_0^\infty [2^{r+1}+(-1)^r/2^{r+1}]/5 z^{r+1} \quad (R > 2)$.

(c) $\sum_0^\infty [(-1)^r/5 \times 2^{r+1} z^{r+1}] - \sum_0^\infty z^r/2^r$.

11.6.3. $1/z + z/6 + 7z^3/360 + \ldots \quad (0 < |z| < \pi)$.

11.6.4. (a) $\tfrac12 - \sum_1^\infty z^r/2^{r+1}, \quad 0 \leqslant |z| < 2; \quad 1 + \sum_0^\infty 2^r/z^{r+1} \quad (|z| > 2)$.

(b) $\sum_0^\infty [(-1)^r 4/(5 \times 3^{r+1}) - 1/(5 \times 2^{r+1})] z^r \quad (0 \leqslant |z| < 2)$;

$\tfrac45 \sum_0^\infty (-1)^r z^r/3^{r+1} + \tfrac15 \sum_0^\infty 2^r/z^{r+1} \quad (2 < |z| < 3)$;

$\tfrac15 \sum_0^\infty [2^r + 4(-1)^r 3^r]/z^{r+1} \quad (|z| > 3)$.

11.6.5. (a) $1/z^2 + 1/3! + z^2/5! - z^4/7!$

(b) $\tfrac12 e^i[-i/t + (\tfrac12 - i) + t(\tfrac12 - i/4) + t^2(\tfrac18 + i/12)]$.

Chapter 12

12.1.1. (a) 3; (b) 2; (c) 2.
12.1.2. (a) 0; (b) $a\exp(-i\pi)$, $a\exp(-i\pi/3)$, $a\exp(\pi3)$, $-a$; (c) $(-1\pm i\sqrt3)/2$; (d) $\pm i$.
12.3.2. Let (a) $f(0) = 0$; (b) $f(-ia) = i/2a$; (c) $f(0) = 1$; (d) $F(-\pi i) = -i/4\pi^3$.
12.3.5. (a) $z = \pm i$, triple, $z = -1$, $\exp i\pi/3$, $\exp(-i\pi/3)$, all double; (b) $z = n\pi$ $(n = 0, \pm1, \pm2, \ldots)$, simple; (c) triple at 0, simple at $n\pi/3$ $(n = \pm1, \pm2, \ldots)$; (d) $(2n+1)\pi/2$ $(n = 0, \pm1, \pm2, \ldots)$, simple; (e) $(2n+1)\pi i/4$ $(n = 0, \pm1, \pm2, \ldots)$, simple; (f) $(2n+1)\pi i/2$, double.
12.3.6. (a) Removable singularity $z = 2$, simple pole $z = -2$; (b) simple pole at $z = \pm2i$; (c) removable singularity at $z = 0$, simple poles $z = n\pi$ $(n = \pm1, \pm2, \ldots)$; (d) removable singularity at $z = 0$, simple poles at $z = 1/n\pi$ $(n = \pm1, \pm2, \ldots)$; (e) essential singularity at $z = 0$; (f) simple poles at $z = (2n+1)\pi i/2b$ $(n = \pm1, \pm2, \ldots)$.

12.4.1. (a) $2\pi i$; (b) 0; (c) 0; (d) $2\pi i c$; (a') πi; (b') 2; (c') 0; (d') $c\pi i - 2b$; (a'') πi; (b'') -2; (c'') 0; (d'') $2b + c\pi i$.

12.5.1. (a) $-i/4$ at $-2i$; (b) $1/\sqrt{(b^2 - 4ac)}$ at $\{-b + \sqrt{(b^2 - 4ac)}\}/2a$, $-1/\sqrt{(b^2 - 4ac)}$ at $\{-b - \sqrt{(b^2 - 4ac)}\}/2a$; (c) -4 at -2; (d) 1 at 0; (e) 1 at $\frac{1}{2}$; (f) $\frac{1}{2}$ at $\frac{1}{2}$; (g) 2 at $\frac{1}{2}$; (h) $\frac{2}{3}$ at -1; $\{1 + \exp(2\pi i/3)\}/3$ at $\exp(-\pi i/3)$, $\{1 + \exp(-2\pi i/3)\}/3$ at $\pi i/3$; (i) $(-1)^n (n\pi + 1)/2$ at $n\pi/2$; (j) $-i/2e$ at i; $ie/2$ at $-i$; (k) $\{\exp(-\pi i/4)\}/4$ at $\exp(i\pi/4)$, $\{\exp(i\pi/4)\}/4$ at $\exp(-i\pi/4)$, $\{\exp(-3i\pi/4)\}/4)$ at $\exp(3\pi i/4)$, $\{\exp(3i\pi/4)\}/4$ at $\exp(-3i\pi/4)$; (l) $-1/a^4$ at 0; $\frac{1}{2}a^4$ at ia; $\frac{1}{2}a^4$ at $(-ia)$; (m) $-i/4a^3$ at ia; $i/4a^3$ at $-ia$; (n) 0 at $-3i$; (o) $\{\exp(-a^2)\}/2ia(b^2 - a^2)^2$ at ia, $i\{\exp(a^2)\}/2a(b^2 - a^2)^2$ at $-ai$, $-i\exp(-ab)\{a^2 - 3b^2 + ab(a^2 - b^2)\}/\{4b^3(a^2 - b^2)^2\}$ at ib, $-i\exp(ab)\{3b^2 - a^2 + ab(a^2 - b^2)\}/\{4b^3(a^2 - b^2)\}$ at $-ib$; (p) $\exp\{12(-1)^n\pi i + 2\sqrt{2}\,i^n(1 + i) + 1\}/16\sqrt{2}(-i)^n(-1 + i)$ at $\sqrt{2}\,i^n(1 + i)$ $(n = 0, 1, 2, 3)$; (q) 1 at -2; (r) $(e - 7e^{-1})/32$ at i, $(-7e + e^{-1})/32$ at $-i$.

12.6.1. E.g. $r = 1/11$.

12.6.3. Zero.

12.7.1. (a) Double pole; (b) regular; (c) essential singularity; (d) regular; (e) regular; (f) regular; (g) regular; (h) simple pole; (i) n-tuple pole.

Chapter 13

13.1.2. Simple poles at ia, $-ia$; double pole at $-b$; double pole at ia if $b = ia$.

13.1.3. (a) (i) 0; (ii) 0; (iii) $\pi/2$; (b) 0 (log 0.4)/6; (c) -2π; (d) (i) $2\pi i$; (ii) 0; (iii) $2\pi i$.

13.2.1. (a) $2\pi \sqrt{(1 - a^2)}$; (b) $2\pi a^n/(1 - a^2)$; (c) $\pi(1 + a^2)/(1 - a^2)$; (d) $2\pi \sqrt{(1 + a^2)}$; (e) $5\pi/8$.

13.3.1. (a) $\pi/16 \sqrt{2}$; (b) $\pi/2 \sqrt{2}\,p^3$; (c) $\pi/2$; (d) $\pi/500$; (e) $3\pi/16q^5$; (f) 0; (g) $\pi/2$; (h) 0; (i) $3\pi/20$.

13.3.4. (a) $1/R^3$ for all amp z; (b) $1/(R^2 - 4)$, amp $z = \pm\pi/2$; (c) $R^3/(R^4 - 16)$, amp $z = \pm\pi/4$, $\pm 3\pi/4$; (d) $R + R^{-1}$, amp $z = \pi/2$, $3\pi/2$; (e) $1/(R^4 - R^2 + 1)$, amp $z = 0$.

13.4.2. (a) $\pi/2e^2$; (b) $\pi/2e$; (c) $(e^2 + 3)\pi/8e^2$; (d) $(1 + m)\pi/4e^m$; (e) π/ae^a; (f) $\pi(e^{-mb}/b - e^{-ma}/a)/(a^2 - b^2)$; (g) $\pi(1 + am)/2a^3 e^{am}$.

13.6.1. (a) $O(R)$; (b) $O(1)$; (c) $O(1)$; (d) $O(R^{-2})$; (e) $O(1)$; (f) $O(R^{-1})$; (g) $O(R)$.

13.7.3. $\frac{1}{2} \tan(k/2)$.

13.8.1. Simple pole at -2; double pole at $\frac{1}{2}$; regular at ∞.

$$P_1 = 9/[25(2 + z)], \qquad P_2 = 39/[50(2z - 1)] + 11/[10(2z - 1)^2].$$

Answers to Miscellaneous Exercises

3. 2, $4i$.

4. $(x + 1) \prod_{r=0}^{3} [x^2 + 2\cos(2r + 1)\pi/9 + 1]$.

5. $e^{ir\pi/3}$, $r = 0, \pm 1, \pm 2, 3$; $2^{1/2}e^{(4n + 3)i\pi/12}$ $(n = 0, 1, 2, 3, 4, 5)$.

7. (i) Circle on join of -3 and $-1/3$ as diameter; i.e. centre $-5/3$, radius $4/3$; (ii) upper unit semicircle; (iii) ellipse, foci 0 and 6, major axis 10.

10. $z = e^{2n\pi i/5}$ $(n = 0, 1, 2, 3, 4, 5)$. $2\cos 2\pi/5$.

11. $a_{2n} = 0$, $a_{2n+1} = (-1)^{n+1} 2^{2n+1}/(2n+1)!$; valid for all x.
$b_n = (-1)^n (n+1) 4^n/\pi^n$; valid $(0 < x < \pi/2)$.

12. $z = 2^{1/6} e^{i(3n+1)\pi/9}$ $(n = 0, \pm 1, \pm 2, 3)$. Roots given by $n = -1, 1, 3$.
$\sqrt{2}\, z^3 = 1 + i\sqrt{3}$.

14. cgt. for $0 < |x| < 2$.   **15.** (a) $-1 \leqslant x < 1$; (b) all real x.

16. (i) (a); (ii) (b); (iii) (a); (iv) (b); (v) (b).

17. $(\pm\sqrt{3}+i)/2, \pm i$.   **18.** (a) $|x| < 1$; (b) $-1 < x \leqslant 1$.

19. (i) cgt; (ii) dgt; (iii) cgt.

23. (a) (i) dgt; (ii) cgt; (c) (i) cgt $|x| < 2$; (ii) dgt.

24. E.g. $\tan x \operatorname{sech}^2 y/(1 + \tan^2 x \tanh^2 y) + i \tanh y \sec^2 x/(1 + \tan^2 x \tanh^2 y)$.

27. 0.0.   **29.** (i) Abs cgt $|z| \leqslant 1$, dgt otherwise; (ii) cgt $\mathcal{R}(z) < 0$; (iii) dgt.

31. (i) $|z| < 1$, abs cgt all k; $|z| > 1$, dgt for all k. If $|z| = 1$, abs cgt $k < 0$; (ii) abs cgt for all k if $\mathcal{I}(z) > 0$; dgt $\mathcal{I}(z) > 0$. If z is real, abs cgt if $k < 0$.

32. $v = (y^2 - x^2)/2 - y/(x^2 + y^2)$, $u + iv = 1/z - iz^2/2$.

33. $w = (z - i)/(z + 1)$; interior of semicircle centre $\frac{1}{2}(1 - i)$, radius $1/\sqrt{2}$; through origin.

38. (i) $2i, -i/2$; (ii) $\zeta = i$; (iv) circles pass through $z = i/2$; (v) $|(z + 2i)/(z - i)| < 2$; exterior of circle centre $2i$, radius 2.

46. $n\pi - 3\pi/8 + (i/4) \log 2$.

47. The quarter plane $\mathcal{R}(w) < 1$, $\mathcal{I}(w) < 0$.

48. E.g. $\sum\limits_{0}^{\infty} [(-1+i)^{r+1} - (1+i)^{r+1}] t^r/2^{r+2}$.

50. $(1/2^{n+1} - 1/3^{n+1})$. Valid $|z + 1| < 2$.

51. $-\pi i/\sqrt{2}$.   **52.** Quarter plane $\mathcal{R}(w) < 1$, $\mathcal{I}(w) < 0$.

53. (a) (i) The half-plane $\mathcal{R}(z) < 0$; (ii) the half-plane $\mathcal{I}(z) < 0$; (iii) the half-line from $-i$ through $-1 - 2i$; (b) (i) the positive real axis; (ii) the upper half-plane.

54. (i) $2\pi\sqrt{(1 - a^2)}$; (ii) $3\pi/4e$.

55. (a) (i) $\sum\limits_{0}^{\infty} -(1/5)(1/4^{r+1} + [-1]^r)z^{2r}$.

(ii) $(-1/5)\left[\sum\limits_{0}^{\infty} z^{2r}/4^{r+1} - \sum\limits_{0}^{\infty} (-1)^r z^{2r-2} \right]$.

(iii) $(1/5) \sum\limits_{1}^{\infty} [4^r - (-1)^r]z^{-2r-2}$.

(b) $1/z^2 - 1/2z + 1/12$.

56. $4\frac{1}{2}$.   **57.** (i) $\pi/6$; (ii) $\pi/6e^3$.

58. (ii) $\sum\limits_{1}^{\infty} [(-1)^r/2^r - 2^r]z^r$; valid $|z| < \frac{1}{2}$.

INDEX

INDEX